Milk Chemistry

Milk Chemistry

– Author –
Rita Khare

BIO - GREEN
BOOKS

Bio-Green Books
New Delhi - 110 002

ISBN: 978-93-90438-87-7 (Hardback)

Published by : **BIO-GREEN BOOKS**
4736/23, Ansari Road, Darya Ganj,
New Delhi - 110 002
Phone: 011-23245578, 23244987
E-mail: biogreenbooks@gmail.com

Printed at **:** Replika Press Pvt. Ltd.

Preface

About 87% of milk is water, in which the other constituents are distributed in various forms. We distinguish among several kinds of distribution according to the type and size of particle present in the liquid.

In milk we find examples of emulsions, colloids, molecular and ionic solutions. An ionic solution is obtained when the forces that hold the ions together in a solid salt are overcome. The dissolved salt breaks up into ions which float freely in the solvent. Thus when common salt sodium chloride is dissolved in water it becomes an ionic solution of free sodium and chloride ions. Ionic solutions are largely of inorganic compounds. In a molecular solution the molecules are only partly, if at all, dissociated into ions. The degree of dissociation represents an equilibrium which is influenced by other substances in the solution and by the pH of the solution. Molecular solutions are usually of organic compounds. In a colloid, one substance is dispersed in another in a finer state than an emulsion but the particle size is larger than that in a true solution. Colloidal systems are classified according to the physical state of the two phases. In a colloid, solid particles consisting of groups of molecules float freely. The particles in a colloid are much smaller than those in a suspension and a colloid is much more stable. An emulsion consists of one immiscible liquid dispersed in another in the form of droplets the disperse phase. The other phase is referred to as the continuous phase. The systems have minimal stability and require the presence of a surface-active or emulsifying agent for stability. In foods, emulsions usually contain oil and water. If water is the continuous phase and oil the disperse phase, it is an oil-in-water (o/w) emulsion, e.g. milk or cream. In the reverse case the emulsion is a water-in-oil (w/o) type, e.g. butter. In summary, an emulsion consists of three elements, the continuous phase, the disperse phase and the emulsifying agent. A dispersion is obtained when particles of a substance are dispersed in a liquid. A suspension consists of solid particles dispersed in a liquid, and the force of gravity can cause them to sink to the bottom or float to the top. For example, fine sand, dispersed in water, soon settles out. An acid is a substance which dissociates to produce hydrogen ions in solution. A base is a substance which produces hydroxyl ions in solution. It can equally be stated that an acid is a substance which donates a proton and a base is a substance which accepts a proton. Buffers are defined as materials that resist a change in pH on addition of acid or alkali. Characteristically they consist of a weak acid or a weak base and its salt. Milk contains a large number of these substances and consequently behaves as a buffer solution. Fresh cows milk has a pH of between 6.7 and 6.5. Values higher than 6.7 denote mastitic milk and values below pH 6.5 denote the presence of colostrum or bacterial deterioration. Because milk is a buffer solution, considerable acid development may occur before the pH changes. A pH lower than 6.5 therefore indicates that considerable acid development has taken place. This is normally due to bacterial activity. The quantities of the main milk constituents can vary considerably depending on the individual animal, its breed, stage of lactation, age and health status. Herd management practices and environmental conditions also influence milk composition. If milk is left to stand, a layer of cream forms on the surface. The cream differs considerably in appearance from the lower layer of skim milk. Proteins are an extremely important class of naturally occurring compounds that are essential to all life processes. They perform a variety of functions in living organisms ranging from providing structure to reproduction. Milk proteins represent one of the greatest contributions of milk to human nutrition. Casein was first separated from milk in 1830, by adding acid to milk, thus establishing its existence as a distinct protein. In 1895 the whey proteins were separated into globulin and albumin fractions. Lactose is the major carbohydrate fraction in milk. It is made up of two sugars, glucose and galactose. The average lactose content of milk varies between 4.7 and 4.9%, though milk from individual cows may vary more. Mastitis reduces lactose secretion.

The present publication deals with composition of milk, milk ingredients analysis, microbiological analysis, adulterants and contaminants analysis. Although several Govt. quality agencies have published their booklets/standards/ official publications and some books are also available on dairy chemistry but such scattered technical information does not serve dairy men as per their requirements.

—Dr. Rita Khare

Contents

1

Known Inherited Characteristics in Dairy Cattle

Introduction

Today's US dairy cows produce twice the milk of cows in 1957. This improvement results from genetic selection for milk yield, improved feeding practices and other management changes. Unfortunately, reproductive function has declined steadily for most of that period, with only a slight, recent recovery. The reasons for the decline in fertility are complex and not completely understood. Inbreeding, lack of selection for reproduction, stress of lactation, and changes in cow environment have all been implicated. This paper discusses two effective genetic solutions to the problem of infertility in dairy cattle: crossbreeding and selection for reproductive traits.

An animal (or a person) often looks a lot like one or both of its parents because of the genes (DNA) that are passed along from the father and mother to the animal. For example, kittens grow up to look like cats, and calves grow up to look like cattle. Genes influence how all the traits of an animal are expressed, sometimes directly and strongly (for example, coat and eye colour, height, or weight) ... and other times indirectly or in combination with other genes (for example, how much milk a cow produces, the composition of the milk, or how well the cow resists disease). Because traits are influenced by genes, the appearance or behaviour of animals can be changed by choosing animals with the features you consider most desirable to be parents of the next generation of animals. Genetic change is permanent and adds up over time.

Genetic Inheritance – Why does a cow look (or not look) Like its Parents?

Gene Expression: Genes work in pairs. Sometimes each gene has an equal influence on how a trait is expressed, but many times one gene is dominant over the other (recessive) gene in determining the trait. For example, whether or not a cow has horns is controlled by a dominant-recessive pair of genes. The gene for no horns (polled) is dominant, and the gene for horns is recessive. A cow with 2 "polled" genes will have no horns. A cow with 2 "horned" genes will have horns. A cow with 1 polled gene and 1 horned gene will not have horns because the horned gene is recessive and the polled gene is dominant.

Genetic Evaluations: Differences between individual animals that can be passed on to future generations (inherited) can be measured through genetic evaluation. How a calf will look or produce can be estimated (scientifically guessed) by looking at an animal's pedigree (ancestor chart) to see how its ancestors looked or produced, When the calf grows up, its own appearance and ability to give milk can be used to provide more information on how its genetics can be passed to its calves. Genetic evaluations are important for choosing parents with superior genes for traits of interest (such as milk production or long life).

Improving dairy cow fertility through genetics:

- Dairy cattle fertility is determined largely by environmental factors but the genetic component should not be ignored.
- The decline in fertility in dairy cattle that has occurred in the last 50 years was caused in part by changes in genetic composition of dairy breeds.
- Crossbreeding can result in an improvement in fertility and longevity and a decrease in milk yield.
- The profitability of crossbreeding will vary from farm to farm; good estimates of the economic value of crossbreeding have not been made.
- A genetic evaluation tool for reproduction now exists in the form of Daughter Pregnancy Rate and one can expect that daughters of bulls with high Daughter Pregnancy Rates will be more fertile that contemporary herd mates.

Crossbreeding

Crossbreeding in dairy cattle is generating renewed interest due to concern over inbreeding. The inbreeding coefficient for US Holsteins and Jerseys has increased steadily from 1960 to 2007 (1960 is the 'base' year for U.S. inbreeding calculations; the inbreeding coefficient for animals born that year is assumed to be zero).

Cattle have about 21,000 functional genes on 30 pairs of chromosomes. As pairs, the chromosomes carry paired (2) copies of each gene– one inherited from the father and one from the mother. Inbreeding increases the likelihood that the two inherited copies of a gene are identical. The negative effects of inbreeding are seen when identical undesirable genes are inherited from both parents. An inbreeding coefficient estimates the percentage of identical genes pairs inherited by an animal. Mating selection should strive to keep inbreeding coefficients of offspring to less than 6.25%. Average inbreeding coefficient for animals born in 2007 was estimated at 5.1% for Holsteins and 7.5% for Jerseys. Thus, many animals in both breeds surpass the 6.25% threshold.

Crossbreeding takes advantage of heterosis (improved performance of offspring over the average performance of parents). Essentially, heterosis results from the opposite of inbreeding depression; chances are increased that gene pairs include a different copy from each parent. This increases the chance that a 'better' version is used by the animal.

Crossbreeding also provides opportunities to exploit superior fertility and health traits of certain dairy breeds. Researchers from the University of Minnesota have been studying the performance of the Holstein crosses of several of these breeds including Normande, Montebeliarde and Scandinavian Red (includes Norwegian Red and Swedish Red breeds).

In these tests, crossbreeding improved fertility and longevity although milk yield decreased. Days open during the first lactation averaged 19-27 days less for the crossbred groups than for purebred Holsteins. In addition, conception rate at first insemination was significantly higher for Normande x Holstein and Montebeliarde x Holstein than for Holstein. In addition to improvements in reproductive traits, reduced stillbirths and increased cow longevity were also associated with crossbreeding. On the other hand, first lactation Holsteins produced 2,706 lb more milk than Normande x Holstein, 1,317 lb more milk than Montebeliarde x Holstein and 1,050 lb more milk than Scandinavian Red x Holstein. The milk yield advantage for purebred Holsteins increased in second lactation: the 305-day milk yield averaged 26,194 lb for Holstein, 21,863 for Normande x Holstein, 23,547 for Montebeliarde x Holstein and 23,683 for Scandinavian Red x Holstein.

Despite its beneficial effects, there are at least two concerns about crossbreeding. The first relates to the loss of heterosis when breeding crossbred offspring. Loss of heterosis can largely be prevented by the use of a three or four-breed sire rotation. Use of a three-breed rotation allows a herd to maintain 86-88% of the heterosis of the first cross and use of a four-breed rotation maintains 93-94% of the maximum heterosis possible. The second concern with crossbreeding is the loss of milk yield when compared to pure Holsteins. Whether the improvements in fertility and health traits caused by crossbreeding compensate for this loss of milk yield will vary by farm and region. Differences in milk price, feed costs, fertility, etc. mean that crossbreeding may be economical for some operations and not for others.

Selection for Reproduction – Daughter Pregnancy Rate

Until recently, there has been little interest in using genetic selection to improve reproduction in dairy cattle. The heritability of reproductive traits is relatively low compared to production traits like milk yield, so genetic improvement is more difficult. Heritability is the proportion (percentage) of the observed variation in a trait that is due to genetics. When heritability is low, it is difficult to make much progress in genetic selection because the probability that the best animal is best for genetic reasons is relatively low. Heritability estimates for reproductive traits are 0.05 or less. In other words, less than 5% of the variation between cows in their reproductive function is caused by differences in genes. This is not surprising because reproduction is greatly affected by non-genetic factors – level of feeding, air temperature, skill of the inseminator, etc. It is a mistake, however, to conclude that low heritability means that there are no genes controlling reproduction. In fact, scientists are now finding specific genes that affect reproduction function.

In 2003, the USDA began estimating the genetic merit of bulls for reproduction. The trait used is Daughter Pregnancy Rate (DPR). This term is calculated from days open and is directly related to the proportion of females eligible to become pregnant in a 21-day period that actually become pregnant (i.e. the 21-day pregnancy rate).

The heritability of DPR is only 0.04 but work is underway to control for other factors affecting days open. This will improve the accuracy of estimates of genetic ability for reproduction. Weigel (2006) reported that the top 10% of Holstein bulls had a DPR 4.9% higher than bulls in the lowest 10%. This difference corresponds to a difference in days open of 20 days (a 1% difference in DPR is equal to a 4 day difference in days open).

Although low heritability makes progress more challenging, reproduction can be improved through genetic selection. The continual decline in DPR for US dairies from 1959 to 1995 was arrested in the latter 1990s and has been increasing since 2000. This increase is likely due to increased management attention to reproduction and the implementation of timed artificial insemination protocols.

Additionally, selection for Productive Life (first evaluated in 1994), which is highly related to reproduction, has likely contributed to recent improvements. Changes in sire and cow breeding value for DPR suggest that the long-term decline in genetic merit for fertility also stopped beginning in the mid-1990s and has since improved slightly.

Alta Genetics, which markets bulls with high breeding values for DPR, has posted data on their website supporting the idea that selection of bulls for DPR can affect cow fertility. In one dairy, fertility of two-year old daughters of Blastoff, the number one ranked Holstein bull for DPR, was compared to fertility of contemporary herd mates. Reproductive function in Blastoff daughters was much better than for contemporary controls. Caution should be used when interpreting these data – number of Blastoff daughters was low and data are not from an independently-conducted, controlled experiment.

Improvements in DPR will pay off economically. De Vries (personal communication) has estimated that a 1-percentage unit increase in 21-day pregnancy rate is worth about $20/cow/year. The value is greater when pregnancy rates are low. The marginal return on 1 lb extra milk achieved by genetic selection is about $0.10/lbs. In other words, improving 21-day pregnancy rate by 1% is equivalent economically to improving milk yield by 200 lb per cow per year.

Take Home Messages:

- Dairy cattle fertility is determined largely by environmental factors but the genetic component should not be ignored.
- The decline in fertility in dairy cattle that has occurred in the last 50 years was caused in part by changes in genetic composition of dairy breeds.
- Crossbreeding can result in an improvement in fertility and longevity and a decrease in milk yield.
- The profitability of crossbreeding will vary from farm to farm; good estimates of the economic value of crossbreeding have not been made.
- A genetic evaluation tool for reproduction now exists in the form of Daughter Pregnancy Rate and one can expect that daughters of bulls with high Daughter Pregnancy Rates will be more fertile that contemporary herd mates.

Inherited Defects in Cattle

Defective calves within our total herd of breeding cattle become important under certain conditions, including heritability of the defect, freqency of occurrence, severity, possibility of cure by nutritional thereapy or surgery, and association with other traits, either deleterious or of economic importance.

With the current viewpoint that all traits are affected by both inheritance and environment (feed, housing, temperature, age, and management), it nevertheless becomes helpful to speak of inherited traits as those predominantly effected by gene action and interaction. Hair and body colours are good illustrations of such traits predominantly conditioned by gene action. Those traits affected to a large extent by any environmental stimulus would be called environmental traits. Highly infectious diseases illustrate this group.

Those traits in each group affected stone what equally, or significantly, by both inheritance and environment might also be put into an intermediate classification, to emphasize the interaction of both sets of forces. Traits requiring high energy transformation from feed to food, such as milk production, stand as a good example of the importance of this interaction.

From the standpoint of basic herd management involving both investment and operating factors, the investment is represented by the genes and gene combinations (genotypes) assembled in the herd; whereas, the proper combination of environment factors becomes the operating forces. The two are combined, each with respect to the other, for management to be successful.

Inherited defects may be defined as those primarily caused by gene action, which impair the usefulness of the animal under normal or usual conditions of husbandry. If a trait is responsible for the failure to survive under such conditions, it falls into the classification of a lethal; regardless of the time after conception that the animal dies, or the unusual care required for survival. About 30 different genes, probably acting in pairs without special consideration of the effect of other genes, cause death either of the embryo, or of the calf shortly after birth. Another 15 genes have been implicated in scientific reports to affect specifically the fertility of either the cow or bull.

An equal number of physical weaknesses and traits discriminated against in the show ring, and type classification programs of the various breed-registry organizations, in the main are presumed to have an inherited basis and, therefore, to be caused or accentuated by gene action.

The list thus accounts for approximately 60 defects affected by inheritance. The prevalence of defects is not well known. However, a survey conducted in cooperation with the Central Ohio Breeding Association gives some pertinent information. Abnormalities falling rather distinctly into 19 different categories were found to include taillessness, wry taft, and other tail abnormalities; blindness and small-eyedness; bull dog, parrot jaws, cleft palate; frozen joints, flexed pasterns, and extra limbs; hairlessness; congenital dropsy; no anus, and those born inside-out. Red in black breeds, which is considered abnormal by some breeders, was also found. Two or more of the above abnormalities have been found to occur in the same animal. Thus, some calves are born without a tail, or anus, or eyeballs.

One problem is to establish the bias in this estimate, and to determine the precise role of inheritance in causing the occurrence of defects. The report that more abnormalities are found in dairy calves than in the young of beef cattle, swine, sheep, or other species is of considerable interest. This conclusion may be realistic, because dairy cattle are subject to closer observation and more adequate records than other domestic animals. The fact that only one is usually born at a time may cause a defective offspring to have mote apparent meaning than in cases where only one in a whole litter is defective.

A propensity for high production might make the birth of a dead or deformed calf seem inconsequential to the main economic reason for keeping the cow. Regardless of this difference between classes of livestock, breeders are concerned, because any inherited trait caused by recessive genes can become disseminated widely through the use of popular bulls that carry the responsible genes. In fact, examples of this association exist in this country, to account for the present frequency of red in black breeds, and in Europe to account for certain lethals being widely disseminated. When daughters of such bulls are bred to others carrying the same gene, one-eighth (121/2%) of the offspring will be afflicted (will be homozygous for the pair of recessive genes responsible).

The elimination from the breeding herds of only cattle showing traits caused by recessive genes actually does but little to control the frequency with which the trait appears. The more rigid selection, that of culling the parents—at]east the sire—of afflicted calves would be effective, but appears to be too drastic to be popular with cattle breeders. By and large, some method of detecting the carrier needs to be found, before much hope can be offered to the breeder, breeding organization, or breedregistry association which seeks to control the frequency of a particular geue and thus to govern to some extent the gene-pool investment.

If genes do not get eliminated with affected cattle, they may remain undetected for many generations. An inherited trait may take on considerable economic importance because the responsible genes can be, and in some cases have been, concentrated in a herd or breed through the selection process. The time when inherited defects show up varies widely. Some perhaps kill the fetus shortly after conception and consequently either go unnoticed or become nmnifested only as an irregular estrous cycle in the dam. Others show up later when the arrested growth or lualdevelopment of certain organs or parts causes death of the fetus and subsequent expulsion. However, more defects show up at birth, or shortly after, than at any other short period of time.

The frequency of a defective trait in a particular herd has to exceed a threshold value before it becomes important in the management of the herd operation. From the standpoint of the investment, however, the occurrence of an inherited defect signals its presence in the gene-pool. If an inherited defect is possessed by a bull whose frozen semen is widely used, an increase in the frequency obviously will follow.

The severity of defects, either anatomical, physiological, or both, varies from slight to extreme. On the extreme side are those causing death in any circumstance of environment, such as in bulldog calves. Slightly less extreme are those permitting survival providing special environment is furnished. Cattle with severe shortening of the lower jaw (parrot jaw) can survive if provided harvested forage or all abundance of long grass to graze.

Severely defective traits handicap the dairyman whose cattle possess them, whether he is a breeder or a commercial milk producer who raises his own replacements. Less severe traits of type are discriminated against in the show ring and are selected against by breeders, because of the higher breeding value of' cattle meeting the type standards.

After inherited causes are established for defects, it may be possible to ameliorate the expression, or even the occurrence, by some treatment such as special diet or surgery. For instance, an operation might permit a calf with imperforate anus to survive. In Holland, surgical treatment of the retractor penis muscles effected a cure, operationally speaking, of bulls involved with impotenia coeundi (inability to copulate). Gihnan recently suggested the possibility of a vitamin A imbalance effectino' the penetrance of genes causing hydrocephalus.

While studies of this sort are, to date, few in nmnber, further investigations may extend the possibility of obtaining maximum economic advantage of cattle possessing genes for certain defects, until the breeder can eliminate the genes from the breeding herd through a longtime, planned culling program.

When one trait is genetically associated with a second trait—either because causative genes are closely linked, i.e., close together on the same chromosome or because of the pleiotropic effect of the same gene(s)—selection for one of the traits, no matter how desirable, would result in increasing the frequency with which the other trait appeared. In cases where the heterozygote for a pair of genes, one of which causes defective offspring in the homozygote, is favoured in selection, there likewise would be built up an increased frequency of the gene.

If the heterozygous calf carrying the gene for dwarfism can be visually detected, and is selected for because of greater compactness of beef form, the gene for dwarfism thus would be favoured in selection. Such selective advantage could account for the greater increase with which dwarf calves now appear in some beef breeds, as compared to dairy breeds, because it is estimated that three cattle generations ago (15 years) the incidence of dwarfism was no greater in beef than in dairy breeds.

What of the Future?

By way of prognosis, it can be predicted that the problem of abnormalities will become greater with the application of new techniques that will come with the passing of time. The first reason is that ionizing radiations is one of the factors increasing the rate at which genes mutate. The estimated number of genes per cow is great enough to suggest that each head of cattle could possess a gene that is different (and probably undesirable) from those of the parents. These mutant genes will be able to pass on to offspring even though they did not come from the parents. An increase in mutant genes can be expected with the increased use of fissionable material in war and in peace.

The second reason is the increased skill in surgery and internal medicine now available to our practicing veterinarians. Their use, along with improved husbandry, results in the saving for parenthood animals that in nature would not survive, or would be placed at a disadvantage in following through with the instinct of mating. Thirdly, breeders sometimes prefer, and therefore select, animals that carry undesirable genes. This illustrates how the frequency of a gene can be inadvertently increased, and that purebred cattle may not always be purebreeding parents.

To what, then, do these things come? How can the problem be resolved? One method is to select sires after they have been proved to be free of genes considered undesirable. Until laboratory methods become available for detecting carriers, proof can be obtained for the absence of single recessive genes expressed in either sex by mating a bull to about 20 daughters or sisters. The accuracy of the proof, however, will depend on the number of such nmtings made, and to obtain the degree of reliability often sought in experimental studies (P——.05), 30 to 35 such matings would be needed. The added expense involved would probably be greater than the value of the added accuracy, unless a particularly troublesome defect was involved. The second method is to learn to live with such genes, but to avoid letting any one undesirable gene become too prevalent in any one herd. If one sire is known to transmit an undesirable gene to half its offspring, the producer should avoid breeding its daughters to another bull carrying the same gene.

The salient point here is that either program includes a knowledge of what genes are involved, and how they react to predispose a cow undesirably. At the present time, this area of cattle breeding is engulfed in abysmal ignorance. Therefore, before any program can be started logically, a research program must be effected to find out for which abnormalities inheritance is of practical importance, and either how the responsible genes can be eliminated, or how the occurrence of the abnormalities can be prevented otherwise. If any one, or all, of the defects presented above is inherited, it is more important to have such information now than at any time in the past.

Breeders presumably want the facts, so that they can safeguard the future of their breeding program. Those responsible for selecting bulls to be used in artificial breeding want to know what their bulls transmit, and how to explain the occurrence of defective calves on the farms of their cooperators. Those buying bulls for their own herds want to know how to breed livestock that will have greater value to themselves and to their customers.

Those responsible for registering livestock to become parents of the next generation areentrusted by their members and other constituent users of their service to improve the genetic potential of the breeds. By virtue of their position of leadership, breed-registry officials, breeding organization managers, and research personnel in our experiment stations have an obligation to point the way to progress. In the active cooperation of all interested organizations and breeders would seem to lie the hope for advancement.

Systems of Breeding

There are three steps in establishing a logical genetic strategy for beef production. First, determine the production and marketing conditions and match applicable levels of animal performance to these conditions.

Second, choose a breeding system. Third, select genetic types, breeds, and individuals within breeds for compatibility with the first two considerations.

Beef cattle producers face two types of decisions concerning breeding systems—which animals are allowed to reproduce and which males are bred to which females.

Mating Plans

Mating plans can be based on: 1) randomness; 2) genetic relationship (pedigree); or 3) performance or visual appearance (phenotype).

- Random mating does not mean random selection. Rather, individuals are selected for breeding. Then they may be managed in one breeding group, with one or multiple sires, or both males and females can be randomly gate-cut into separate breeding groups. Either way, there is no action taken to determine which animals mate. Random mating is a rather common procedure, especially for multiple-sire herds where it is difficult to maintain more than one breeding group.
- Pedigree mating implies that all individuals in a genetic population (such as a herd, family line, or breed) are related to some extent. One pedigree plan mates individuals more closely related than the average of the population; it is termed inbreeding. While long-term inbreeding in a closed herd may increase genetic uniformity, inbreeding usually reduces performance, especially in fertility and survival. This is called inbreeding depression. One type of inbreeding is line breeding, which is used to concentrate the genetic influence of some line or individual while minimizing increases in inbreeding. Mating animals less related

than average is called out breeding or out crossing. Out crossing of lines within a breed can restore performance lost to inbreeding depression. Mating individuals of different breeds is called crossbreeding, which often increases performance above what might be expected from the parent breeds. This effect is called hybrid vigour or heterosis. It is commonly thought that out breeding increases variability, but well-planned out crossing or crossbreeding produces uniform progeny.

- Phenotype mating plans are based on performance or visual appearance, not pedigree, and are called assortative. Mating individuals most alike in performance or appearance is positive assortative mating such as mating the heaviest males to the heaviest females or the shortest males to the shortest females. Compared to random mating, this results in more variation in progeny, fewer progeny near average, and more extremes.

This plan is used mainly in hopes of producing a few extreme animals to quickly change a population. Positive assortative mating is sometimes called "mating the best to the best," a sound concept if parents are superior in all important factors.

Examples of the opposite plan, negative assortative, are mating the heaviest males to the lightest females or the shortest males to the tallest females. Consequences of this scheme, compared to random mating, are decreased variation, more individuals near average, and fewer extremes. If population-average performance in offspring is optimum, then this plan is useful. Often these types of matings are used to correct problems. For example, in a herd with milk production levels too high for existing forage resources, sires of lower milking genetics would produce better adapted replacement heifers. Unless dramatic genetic change is needed, negative assortative mating often is a sound strategy.

Crossbreeding

Crossbreeding begins with the mating of two purebreeds. The term F1 applies to progeny of such a cross. A more useful definition of F1 is the progeny of parents with no common genetic background. The most desirable crossbreds are results of genetically superior purebred parents. In fact, superior purebreds may easily exceed the performance of crosses from mediocre purebred parents.

There are three benefits of crossbreeding over restriction to a single breed (straightbreeding)—heterosis, breed combination, and complementarity.

Heterosis

Heterosis is measured as performance of crossbred progeny compared to the average of purebred parents. Heterosis is usually positive.

It is highest in the progeny of least related parents. For instance, there is greater heterosis in crossing the genetically dissimilar Hereford and Brahman breeds than in crossing the more similar Hereford and Angus.

Heterosis is reduced when the same breed is a constituent of both parents. As an example, if cattle sired by Angus and out of Hereford are bred back to one of these breeds (a backcross), the resulting offspring average 50 percent less heterosis than the F1 Angus-Hereford.

If the F1 is bred instead to a third breed, then heterosis of progeny can be either higher, the same, or lower, depending on the genetic relationship of the third breed to Angus and Hereford. If, instead of a backcross, you mate two F1s of the same breed makeup, the progeny, called F2, also average 50 percent reduction of heterosis from the F1, the same as a backcross. But if you intermate those F2s, producing an F3, there is no additional loss of heterosis, on the average, beyond that experienced in going from the F1 to F2. Heterosis is reduced beyond the F3 only to the extent that inbreeding occurs.

Characteristics differ in heterosis. Heterosis is highest in fitness traits such as fertility, livability, and longevity. It is intermediate in milk production, weight gain, feed efficiency, and body size. It is lowest in carcass traits. Heterosis is highest in factors affecting efficiency in dams.

Breed Combination

Even if heterosis was not a factor, there could be benefits merely from combining breeds with different characteristics to produce a superior package. For example, females with genetics for high carcass quality but small body size and low rate of gain could be mated to sires with genetics for large size and fast weight gain but low carcass quality, resulting in progeny acceptable in both growth and carcass quality. In many instances, favourable combinations are the most important benefit from crossbreeding.

Complementarity

The mating just discussed might be called complementary, as it combines parents with differing strengths and weaknesses to produce desirable progeny. However, what if the females in the example were as large in body size as could be efficiently maintained on that particular forage resource? The smaller body size of these females is an advantage for cow adaptability in this situation but a disadvantage in gaining ability of progeny. That disadvantage could be countered with large, fast-gaining sires. But the heifers from that mating would not be useful for replacements in that herd, as they would be too large in body size. The only way to exploit this mating in that environment is to continually use a particular genetic type of female and a different type of sire.

This technique is called complementarity, and it is possible only with a particular breeding system, discussed below.

Types of Breeding Systems

There are two basic breeding systems. If the source of replacement females is heifers produced in the herd, there is a continuous system. If heifers are not put back in the herd, there is a terminal system. Differences in these systems must be well understood, or serious mistakes can be made.

Continuous

A continuous system produces its replacement females but requires an external infusion of sires (unless inbreeding is involved, and that is rarely desirable in commercial production). Since replacement females are retained in this system, the cowherd has genetics of both the sires and dams. Therefore, if sires have traits that are undesirable in brood cows, those traits cannot be hidden in a continuous system. Both sires and dams in continuous systems should be similar in important traits and without any undesirable characteristics. Genetic extremes generally are not compatible with continuous breeding systems.

Terminal

In a terminal system, both replacement females and sires must come from external sources; they are either purchased or come from another herd. However, since heifers produced in terminals are not retained for breeding, there is more flexibility in choice of genetic types. Specialized maternal and sire types can be used in terminals, since undesirable traits can be masked in a properly designed system.

A combination of relatively small dams bred to larger sires in a terminal system fully exploits complementarity. However, in some cases, breeds similar in body size also are useful for terminals as, for example, where climate favours females of heat-tolerant breeds, many of which are relatively low in carcass quality. Sires from breeds known for high carcass quality, most of which are no larger than medium in size, might be the best choice in this case. Some complementarity in body size and weight gain is given up for female adaptability.

Continuous Systems Straightbreeding

Here the same breed of sire and dam is used continually, so progeny usually are rather uniform in appearance. Straightbreeding is particularly useful in producing parents for crossbreeding. The biggest shortcoming of commercial straightbreeding is the important lack of heterosis.

True Rotations

True rotation systems use two or more breeds and the same number of breeding groups. The simplest rotation is a two-breed, sometimes called a crisscross. A different breed of sire is used continually in each of the two breeding groups. Replacement heifers are moved or rotated for breeding from the group where they were produced to the other group, where they remain for all of their lifetime matings.

In a rotation of three or more breeds, a heifer is placed in the breeding group with the breed of sire to which the heifer is least related. This ensures minimal loss of heterosis in progeny.

Because they require multiple breeding groups, true rotations are rather complicated unless artificial insemination is used. (A. I. simplifies many of the mechanics of most crossbreeding systems.) Once a true rotation is fully in place, all breeding groups are present every year. Also, a compromise must be made between complementary matings and uniformity between groups. You cannot maximize both in rotations. Because of these complexities and limitations, true rotations are uncommon.

Sire Rotations

Sire rotations are sometimes called rotations in time. Instead of rotating females among multiple breeding groups as in true rotations, sire breeds are changed periodically in a single breeding group. A sire breed might be used for from one to several breeding seasons, most commonly for two or three. Ordinarily, a single breed of sire is used during a breeding season to produce more uniform progeny and simplify identification of the breed composition of potential replacement females.

Heterosis is lower in sire rotations than in true rotations, though the reduction is slight in well planned systems. Highest heterosis is maintained by keeping replacement heifers out of dams that are least related to the heifer's breed of sire. This merely requires identifying a dam's breed of sire, if a single breed of sire is used in a breeding season.

Sire rotations are much simpler to conduct than true rotations, because there is only one breeding group. This is one of the most common crossbreeding systems. Unfortunately, in many cases such plans are conducted haphazardly, with little thought given to a logical schedule.

Terminal Systems

Static Terminal

In a static terminal, replacement females must come from outside, either by purchase or from another herd. It is simplest to purchase replacement females because then only one breeding group is needed for the terminal cross. This is a particularly simple plan when purchases are limited to females that have calved at least once or twice, in which case there are no heifers that require separate facilities and easy-calving sires.

A straightbred terminal is mechanically possible, but there usually is no good reason to do so because the benefits of crossbreeding are absent. A possible exception is if a strong market exists for some straightbred and the breeder does not wish to or cannot develop heifers.

A two-breed static system, using straightbred males and sraightbred females of different breeds, produces heterosis in crossbred calves. However, such a system forfeits the considerable benefits of heterosis of crossbred dams.

A three-breed terminal is more efficient. It uses two-breed F1 cows and a third breed of sire. First, straightbred females with desirable maternal traits are produced. Then these are crossed with another desirable maternal breed to produce the F1. Then the F1 females are used in a terminal cross. In a complete static system, about one-fourth of the females are straightbred, about one-fourth produce the F1, and only about one-half of the females are in the terminal portion. Someone must perform all these functions in order for three-breed terminals to be possible, and this requires time and expense. The unique advantage of static terminal crossing is the opportunity to fully exploit complementarity. The main disadvantage is in the creation of replacement females.

Rotation Terminals

A rotation terminal (actually a combination of the two basic systems) is designed to solve some of the problems of providing replacement females for static terminals. Here, a rotation system produces replacement females both to keep itself going and for use in a separate terminal. In most instances, middle-aged females (4 to 6 years old) are moved out of the rotation to the terminal, because they are less prone to calving problems if terminal sires are large in body size.

For a rotation terminal, only two breeding groups are needed—one for the sire rotation and one for the terminal cross.

Heterosis is relatively high in these rotation terminals, because all progeny and breeding females are crossbred. However, a high percentage of the rotation heifer progeny must be retained for replacements, so there is little opportunity for selection of females. Approximately 65 to 75 percent of sale calves are from the terminal, with most of the rest being male calves from the rotation.

Composites

A composite is formed from two or more established breeds, usually in exact percentages that can vary depending on the goals. There is specific attention given to retaining heterosis as generations progress. The primary motivation for creating composites is to create desirable breed combinations while producing some heterosis without continual crossbreeding.

Composites as discussed above are not breeds in the usual sense of the word. There are numerous breeds that have been created by combining existing breeds. Formula breeds contain specific percentages of the constituent breeds. Pool breeds do not have specified percentages. These combination breeds also retain some heterosis, but that is not usually a primary motivation in their creation or propagation.

For a more complete discussion of this subject, see another publication in this series, E-180, "Texas Adapted Genetic Strategies for Beef Cattle—VI: Creating Breeds and Composites."

Breeding Systems and Breeding Groups

The choice of breeding systems depends partly on the number of separate breeding groups that can be maintained. The development, breeding, and calving of heifers is conducted most efficiently in a management group separate from older females using easy-calving sires.

One Breeding Group

One-breeding-group herds, ranging from those requiring only one bull to large, multiple-sire herds, have several choices of breeding systems. Straightbreeding is an option, which could be done with either a traditional or combination breed. A static terminal cross could be run, with F1 females being purchased. A sire rotation could be implemented, using breeds that are similar in functional characteristics. A fourth option for one breeding group is the use of a composite.

Two Breeding Groups

Two groups offer other choices including:

- True two-breed rotation
- Straightbreeding in one group to produce females for use in another group, particularly to create F1 replacement females
- Straightbreeding in one group to produce females for a twobreed static terminal cross in another group.
    - o Purchasing straightbred females for creation of an F1 in one group to be used in a three-breed static terminal cross in another group
    - o Sire rotation in one group, producing replacement females for a terminal cross in another group.

Three Breeding Groups

There are three options that require three breeding groups. One is a true three-breed rotation. Another is a true two-breed rotation generating replacement females for a terminal cross. The third is to carry out all three matings for a complete three-breed static terminal cross (production of straightbred females, creation of F1 females, and the terminal cross).

Multiple breeding groups are more complex to manage, and for each breeding group there is a different breed composition in market animals. This can reduce marketing flexibility. Also, some breed combinations may be less valuable than others. Consider these factors before implementing systems requiring multiple breeding groups.

Efficiencies of Breeding Systems

To compare breeding systems at the cow-calf level, a simple measure of production efficiency is pounds of calf weaned per cow exposed to breeding, which combines reproductive efficiency and calf weight. Values shown are percentage increases above continuous straightbreeding. These increases are due to average levels of heterosis and any progeny weight increase from large terminal sires.

Most of the more complicated plans (true rotations, terminals, and combinations) increase efficiency about 15 percent to more than 25 percent. These estimates are for systems using British and Continental breeds in temperate environments. In harsh tropical or subtropical environments, including tropical-adapted breed types can produce even greater increases.

These are significant advantages over straight breeding. In choosing a breeding system, possible effects on the major profit factors should be considered, including:

- Number of animals to sell
- Pounds per animal
- Price per pound
- Total cost of production.

The measure of efficiency used, pounds of calf per cow exposed, lacks any consideration of animal numbers. Larger cows may wean more pounds per cow. But fewer large cows can be run on the same piece of land, so the number of sale calves is reduced.Pounds weaned per cow does not take into account price per pound. Some breed combinations typically receive price discounts, some severe. Also, heavier calves bring less per pound. Finally, pounds of calf per cow exposed does not consider cost of production.

If high levels of reproduction and calf weight increase costs (particularly nutrition costs), the advantage of crossbreeding may be reduced.

It is a major challenge for beef cattle producers to select breeding systems and breeds compatible with climate, forage conditions, general management practices, and market demands. For a discussion of genetic types and breeds of cattle, see another publication in this series, E-190, "Texas Adapted Genetic Strategies for Beef Cattle—V:

Type and Breed Characteristics and Uses

When selecting a breeding system, give careful thought to the entire process. Do not embark on the first stage of a system without planning for subsequent stages. A system that works well for one producer might be completely unsuitable for another.

The following are some of the mating systems generally recognized in animals:

- Monogamy: One male and one female have an exclusive mating relationship. The term "pair bonding" often implies this. It could also mean one or more animals have an exclusive relationship with one or more opposite-sex animals.
- Polygamy: Three types are recognised:
  - o Polygyny (the most common polygamous mating system in vertebrates so far studied): One male has an exclusive relationship with two or more females
  - o Polyandry: One female has an exclusive relationship with two or more males
  - o Polygynandry: Two or more males have an exclusive relationship with two or more females; the numbers of males and females need not be equal, and in vertebrate species studied so far, the number of males is usually less.
- Promiscuity: Any male within the social group mates with any female.

These mating relationships may or may not be associated with social relationships, in which the sexual partners stay together to become parenting partners. As the alternative term "pair bonding" implies, this is usual in monogamy. In many polyandrous systems, the males and the female stay together to rear the young.

In polygynous systems where the number of females paired with each male is low, the male will often stay with one female to help rear the young, while the other females rear their young on their own. In polygynandry, each of the males may assist one female; if all adults help rear all the young, the system is more usually called "communal breeding". In highly polygynous systems, and in promiscuous systems, paternal care of young is rare, or there may be no parental care at all.

It is important to realize that these descriptions are idealized, and that the social partnerships are often easier to observe than the mating relationships. In particular:

- the relationships are rarely exclusive for all individuals in a species. DNA fingerprinting studies have shown that even in pair-bonding, matings outside the pair (extra-pair copulations) occur with fair frequency, and a significant minority of offspring result from them.
- some species show different mating systems in different circumstances, for example in different parts of their geographical range, or under different conditions of food availability
- mixtures of the simple systems described above may occur.

How to Breed Dairy Cattle

A good breeding program is essential to any dairy farm for the simple reason that milk production is dependent on how many cows you have lactating at one time. Remember, too, that although people buy milk year-round, it's unnatural for cows to calve in any season but one. Follow these steps to be sure you're making money every month of the year.

Difficulty: Moderately Challenging

Instructions

Step 1: Learn the cows' reproductive cycle and decide whether to use hormones to time reproduction and increase milk production. This decision determines whether you're a traditional or an organic dairy farmer, and it has profound effects on your breeding operation. Organic dairy farmers can charge twice as much for milk, but traditional farmers produce a great deal more.

Step 2: Supplement cows' feed during pregnancy and, of course, lactation. For cows treated with hormones, this is of special concern because they reach a point where milk production is so high that they simply can't consume enough calories in a day to maintain their weight. It is therefore necessary to ensure that they are nutritionally sound before that point, during pregnancy and in the first phase of lactation.

Step 3: Use artificial insemination if you're a beginner, as dairy bulls are the most difficult to deal with. Breed cows for conformation (femininity, including pelvis size), longevity and milk quality and quantity. The less often you must replace cows, the better.

Step 4: Separate cows due to calve and watch them closely. Provide shelter from the elements—calving stalls in the barn in cold climates and shaded pasture areas in hot climates. Studies have shown that heat stress during parturition plays a role in milk production later, not only for the cow giving birth, but for the calf, if it's a heifer.

Step 5: Bottle-feed calves one gallon of colostrum (ideally) 30 minutes after birth, but certainly within 24 hours. Feed calves pasteurized waste milk with excess colostrum or milk replacers at a rate of 1 gallon per day and supplement with a commercial calf starter feed. Wean them at 5 to 6 weeks.

Step 6: Adjust the diet of your heifer calves so that they're large enough to breed at 14 to 16 months of age, and build your herd with quality heifers born on your own farm to keep replacement costs low.

2

The Purebred Business

Purebreds, also called purebreeds, are cultivated varieties or cultivars of an animal species, achieved through the process of selective breeding. When the lineage of a purebred animal is recorded, that animal is said to be pedigreed.

The term purebred is occasionally confused with the proper noun Thoroughbred, which refers exclusively to a specific breed of horse, one of the first breeds for which a written national stud book was created since the 18th century. Thus a purebred animal should never be called a "thoroughbred" unless the animal actually is a registered Thoroughbred horse.

True Breeding

In the world of animal breeding, to "breed true" means that specimens of an animal breed will breed true-to-type when mated like-to like; that is, that the progeny of any two individuals in the same breed will show consistent, replicable and predictable characteristics. A puppy from two purebred dogs of the same breed, for example, will exhibit the traits of its parents, and not the traits of all breeds in the subject breed's ancestry.

However, over time, there are also concerns that breeding from too small a gene pool can lead to inbreeding and the development of negative characteristics or even a collapse of a breed population due to inbreeding depression. Hence, there is continuing tension within many purebred animal breeds over the question of when a breed may need to allow "outside" blood in for the purpose of improving the overall health and vigour of an animal breed.

Because pure-breeding creates a limited gene pool, purebred animal breeds are also susceptible to a wide range of congenital health problems. This problem is especially prevalent in competitive dog breeding and dog show circles due to the singular emphasis on aesthetics rather than health or function. Such problems also occur within certain segments of the horse industry for similar reasons. The problem is further compounded when breeders practice inbreeding. The opposite effect to that of the restricted gene pool caused by pure-breeding is known as hybrid vigour, which generally results in healthier animals.

Pedigrees

A pedigreed animal is one that has its ancestry recorded. Often this is tracked by a major registry. The number of generations required varies from breed to breed, but all pedigreed animals have papers from the registering body that attest to their ancestry.

The word "pedigree" appeared in the English language in 1410 as "pee de Grewe", "pedegrewe" or "pedegru", each of those words being borrowed to the Middle French "pié de grue", meaning "crane foot". This comes from a visual analogy between the trace of the bird's foot and the three lines used in the English official registers to show the ramifications of a genealogical tree.

Sometimes the word purebred is used synonymously with pedigreed, but purebred refers to the animal having a known ancestry, and pedigree refers to the written record of breeding. Not all purebred animals have their lineage in written form. For example, until the 20th century, the Bedouin people of the Arabian peninsula only recorded the ancestry of their Arabian horses via an oral tradition, supported by the swearing of religiously-based oaths as to the asil or "pure" breeding of the animal. Conversely, some animals may have a recorded pedigree or even a registry, but not be considered "purebred". Today the modern Anglo-Arabian horse, a cross of Thoroughbred and Arabian bloodlines, is considered such a case.

Purebreds by Animal

Purebred Dogs

A purebred dog is a dog of a modern breed of dog, with written documentation showing the individual purebred dog's descent from its breed's foundation stock. In dogs, the term breed is used two ways: loosely, to refer to dog types or landraces of dog (also called natural breeds or ancient breeds); or more precisely, to refer to modern breeds of dog, which are documented so as to be known to be descended from specific ancestors, that closely resemble others of their breed in appearance, movement, way of working and other characters; and that reproduce with offspring closely resembling each other and their parents. Purebred dogs are breeds in the second sense.

New breeds of dog are constantly being created, and there are many websites for new breed associations and breed clubs offering legitimate registrations for new or rare breeds. When dogs of a new breed are "visibly similar in most characteristics" and have reliable documented descent from a "known and designated foundation stock" they can then be considered members of a breed, and, if an individual dog is documented and registered, it can be called purebred.

Purebred Horses

According to the "Four Foundations" theory, the evolution of the horse ultimately produced horses of four basic body types, adapted to different environments. Beginning with the may have been bred true to original type by humans, though emphasizing certain inherent traits (such as a good temperament, suitable to training by humans) to a greater degree than others. In other cases, horses of different body types were cross bred until a desired characteristic was achieved and bred true.

Written and oral histories of various animals or pedigrees of certain types of horses have been kept throughout history, though breed registry stud books trace only to about the 13th century, at least in Europe, when pedigrees were tracked in writing, and the practice of declaring a type of horse to be a breed or a purebred became more widespread.

Certain horse breeds, such as the Andalusian horse and the Arabian horse, are claimed by aficionados of the respective breeds to be ancient, near-pure descendants from one or the other of the original four wild prototypes, though absent a full mapping of the horse genome and other DNA research, such claims are difficult to prove or disprove.

Purebred Cats

A cat whose ancestry is formally registered is called a purebred cat, or a pedigreed cat. A purebred cat is one whose ancestry contains only individuals of the same breed. A pedigreed cat is one whose ancestry is recorded, but may have ancestors of different breeds.

The list of cat breeds is quite large: most cat registries actually recognize between 30 and 40 breeds of cats, and several more are in development, with one or more new breeds being recognized each year on average, having distinct features and heritage. Owners and breeders compete in cat shows to see whose animal bears the closest resemblance to the "ideal" definition, based on breed type and the breed standard for each breed.

Because of common crossbreeding in populated areas, many cats are simply identified as belonging to the homogeneous breeds of domestic long hair and domestic short hair, depending on their type of fur.

Some original cat breeds that have a distinct phenotype that is the main type occurring naturally as the dominant domesticated cat type in their region of origin are sometimes considered as subspecies and in the past received names as such, although this is no longer supported by feline biologists. Some of these cat breeds (with their invalid scientific names for historical interest) are:

- F. catus anura-*the Manx*
- F. catus siamensis-*the Siamese*
- F. catus cartusenensis-*the Chartreux*
- F. catus angorensis-*the Turkish Angora.*

Purebred Livestock

Most domesticated farm animals also have true-breeding breeds and breed registries, particularly cattle, sheep, goats, rabbits, and pigs. While animals bred strictly for market sale are not always purebreds, or if purebred may not be registered, most livestock producers value the presence of purebred genetic stock for the consistency of traits such animals provide. It is common for a farm's male breeding stock in particular to be of purebred, pedigreed lines.

In cattle, some breeders associations make a difference between "purebred" and "full blood". Full blood cattle are fully pedigreed animals, where every ancestor is registered in the herdbook and shows the typical characteristics of the breed. Purebred are those animals that have been bred-up to purebred status as a result of using full blood animals to cross with an animal of another breed. The breeders association rules the percentage of fullblood genetics required for an animal to be considered purebred, usually above 87.5%.

Artificial breeding via artificial insemination or embryo transfer is often used in sheep and cattle breeding to quickly expand, or improve purebred herds. Embryo transfer techniques allow top quality female livestock to have a greater influence on the genetic advancement of a herd or flock in much the same way that artificial insemination has allowed greater use of superior sires.

Wild Species, Landraces and Purebred Species

Breeders of purebred domesticated species discourage crossbreeding with wild species, unless a deliberate decision is made to incorporate a trait of a wild ancestor back into a given breed or strain. Wild populations of animals and plants have evolved naturally over millions of years through a process of natural selection in contrast to human controlled Selective breeding or artificial selection for desirable traits from the human point of view. Normally, these two methods of reproduction operate independently of one another. However, an intermediate form of selective breeding, wherein animals or plants are bred by humans, but with an eye to adaptation to natural region-specific conditions and an acceptance of natural selection to weed out undesirable traits, created many ancient domesticated breeds or types now known as landraces.

Many times, domesticated species live in or near areas which also still hold naturally evolved, region-specific wild ancestor species and subspecies. In some cases, a domesticated species of plant or animal may become feral, living wild. Other times, a wild species will come into an area inhabited by a domesticated species. Some of these situations lead to the creation of hybridized plants or animals, a cross between the native species and a domesticated one.

This type of crossbreeding, termed genetic pollution by those who are concerned about preserving the genetic base of the wild species, has become a major concern. Hybridization is also a concern to the breeders of purebred species as well, particularly if the gene pool is small and if such crossbreeding or hybridization threatens the genetic base of the domesticated purebred population. The concern with genetic pollution of a wild population is that hybridized animals and plants may not be as genetically strong as naturally evolved region specific wild ancestors wildlife which can survive without human husbandry and have high immunity to natural diseases. The concern of purebred breeders with wildlife hybridizing a domesticated species is that it can coarsen or degrade the specific qualities of a breed developed for a specific purpose, sometimes over many generations. Thus, both purebred breeders and wildlife biologists share a common interest in preventing accidental hybridization.

Dairy Crossbreeding: Results from New Research and Information Still Needed

Results from New Research

Three crossbreeding projects are currently under way with dairy cattle in the United States, while results from a fourth trial using data from commercial herds in California have been reported in recent scientific literature. There are additional efforts in other countries, some of which involve intensive grazing management programs. The domestic research referred to here is being conducted in typical U.S. confinement systems.

- Holstein-Jersey crosses versus. purebred Holsteins in Minnesota – oldest crosses have completed third lactation, but the only results are in abstract form as this is updated in June 2007.
- A "backcross" trial at Wisconsin, where over 160 Holsteins were bred to crossbred Holstein-Jersey bulls in AI – first calves were born in 2004 and initial performance results are now being analysed.

- A designed trial comparing purebred Holsteins, Jerseys, and their reciprocal crosses in the Virginia Tech and Kentucky university herds. All calves in this trial have been born, but many are still in calf pens at this writing. Preliminary results reported here are early calf hood traits and a summary of yield data through May 2007. First calving dates for project animals were mid summer 2005 at Virginia Tech and mid winter 2006 for Kentucky.

California Commercial Herd Results

Table shows production for purebred Holsteins and crosses of Normande (a French dairy breed), Montbeliarde (a French dairy breed), and Scandinavian Red bulls on purebred Holstein dams. Highest production was for purebred Holsteins, but milk yields from the Montbeliarde and Scandinavian Red crosses were close to Holstein, with higher components.

Combined fat and protein volume (not shown in the table) for Scandinavian Red-Holstein crosses was not statistically lower than for purebred Holstein cows. These results show that some crossbred cows will produce yields similar to purebred Holsteins, particularly on a component basis. The closeness of performance for yield traits suggests that even small advantages in fitness and fertility will make crossbreds appealing to many commercial producers.

Table also includes calving records for purebred and crossbred cows. Dystocia (calving difficulty) and stillbirths (calf mortality) were greatest for purebred Holstein mothers. Crosses to Montbeliarde and Scandinavian Red sires produced dams with significantly less dystocia and stillbirths than purebred Holstein dams. Normande-Holstein crosses did not differ from purebred Holsteins.

Table: *California trial: comparisons of yield and calving performance.*

		Breed of Cow		
	Holstein	***Normande -Holstein***	***Montbeliarde -Holstein***	***Scand Red-Holstein***
No. Cows in Milk	380	245	494	328
Milk, lb	21,510	18,805*	20,196*	20,460*
Fat %	3.55	3.74	3.65	3.66
Protein %	3.13	3.25	3.20	3.20
No. of Calvings	676	262	370	264
% Calving Difficulty	17.7	11.6*	7.2*	3.7*
% Stillbirths	14.0	9.9	6.2*	5.1*

*Crosses differed from Holsteins ($P < 0.05$). The paper reported volume of components, so component percentages were not tested for significance.

Average dystocia and stillbirth rates are from first calvings when cows were bred to Montbeliarde, Brown Swiss, or Scandinavian Red bulls.

Table shows calving difficulty and stillbirth results by breed of sire when used on first calf Holstein dams. Scandinavian Red bulls produce significantly less dystocia and stillbirths than Holstein sires, and Brown Swiss bulls produce less dystocia than Holsteins.

The advantages of easier calving for matings to Scandinavian Red bulls are clear from these results. Selection programs in the Scandinavian countries have emphasized reduction in dystocia and stillbirths for about 30 years. Many dairy producers across the United States are using Scandinavian Reds on Holstein heifers because of calving ease. All the calves by non-Holstein breeds would have the benefit of heterosis in survival.

We don't have purebred Montbeliarde or Scandinavian Red cows to allow us to estimate heterosis in these calves. What we do know from this trial is that calves sired by Scandinavian Red bulls out of first calf Holsteins have significantly less calving difficulty than calves sired by Holsteins.

Table: *California trial: performance of breed of sire when used on first calf Holstein dams.*

	Breed of Sire			
	Holstein	***Montbeliarde***	***Brown Swiss***	***Scand. Red***
No. of Calvings	371	158	209	855
% Calving Difficulty	16.4	11.6	12.5*	5.5*
% Stillbirths	15.1	12.7	11.6	7.7*

*Different from Holsteins ($P < 0.05$).

Table includes survival and fertility data for purebred Holsteins and the three crossbred breed groups. Crossbreds were more likely to survive through 305 days of first lactation than were the purebred Holsteins. There were no statistically significant differences between the 4 breed groups for days open, but first service conception rate was significantly higher for crossbreds than for purebred Holsteins. The most fertile breed group was the Normande-Holstein cross, which, interestingly enough, was the lowest production breed combination in table.

There is an important lesson in this observation – high production and high fertility are hard to accomplish together. There is a genetic antagonism between these two traits, and some trade-offs are likely as we strive to improve fertility in high-producing dairy cows. Do we have to accept lower production to achieve acceptable fertility? Hopefully not, but we need to develop dairy cows that regain energy balance and retain enough body tissue reserves while making a lot of milk to breed back in a timely manner.

***Table:** California trial: survival and reproduction by breed combination of first lactation cows.*

	Holstein	*Normande -Holstein*	*Montbeliarde-Holstein*	*Scand. Red-Holstein*
No. of Cows	523	363	229	190
% Surviving to 305 Days	86	93*	92*	93*
No. Cows for Days Open	520	375	371	257
Average Days Open[1]	150	123*	131*	129*
No. of Cows for Conception Rate	536	379	375	261
First Service Conception Rate (%)	22	35*	31*	30

*Different from Holsteins ($P < 0.05$).

[1]First calf heifers that had a subsequent calving or had pregnancy status confirmed by a veterinarian: at least 250 DIM was required.

The California trial is the first to compare purebred Holsteins to crosses of some European dairy breeds. It gives us important new information but needs to be interpreted with some caution. The herds that participated made a considered decision to move away from purebred Holsteins. The crosses themselves were novel animals. Consequently, the Holsteins and crosses may have been treated somewhat differently than will be the case in herds that repeat such matings in the future. The results are based on several hundred animals, not many thousands of animals, such as are used to evaluate merit of U.S. pure breeds. Finally, the traits studied are those expressed relatively early in life. Some of the important questions about crossbred-purebred performance relate to performance in mature animals. We have more to learn about these European breeds under U.S. management conditions.

Holstein-Jersey Crosses at Kentucky and Virginia Tech

The Holstein-Jersey crossbreeding project at Virginia Tech and the University of Kentucky was started in 2002. First calves were born in 2003, and first calvings for project animals were in June 2005. Data on early calfhood performance were recently completed, but have not been summarized. Results below are from a preliminary analysis. Table compares the four breed groups in this project for birth weights and dystocia. No significant differences between calves born to the four breed groups were found for stillbirths, so those results are not shown. The breed of the sire appears first in breed group designations. Table shows that birth weights differed for all four breed groups, with purebred Holsteins producing the largest calves, as expected. Jersey-sired calves out of

Holstein dams were larger than Holstein-sired calves out of Jersey dams, suggesting an important effect of breed of dam on birth weights. Dystocia scores were highest for calves sired by Holstein bulls. Jersey dams had as much difficulty giving birth to Holstein-sired calves (the HJ group) as did Holstein dams (the HH group). Conversely, Holstein mothers were equally good "easy calvers" as the Jersey dams when Jersey bulls sired the calves they carried.

Table: *Comparisons of 414 birth weights and 421 dystocia scores by breed group in the Virginia Tech-Kentucky crossbreeding study.*[1]

	Breed Group of Calf			
	HH	***HJ***	***JH***	***JJ***
Birth Weights (lb)	88[a]	65[b]	69[c]	50[d]
Dystocia (1 to 5 scale)	1.7[a]	1.6[a]	1.2[b]	1.2[b]

[1]Stillbirth percentages did not differ by breed group of calf.

[a, b, c, d]Means with different superscripts are statistically different ($P < 0.05$).

Analysis of Dystocia and Stillbirths Using Odds Ratios

Results using odds ratios are in Table. Holstein and Jersey genes for dystocia or stillbirths have one set of effects on the calf but a separate effect from the mother. These are called "additive" and "maternal" effects. A third effect, heterosis, results from combinations of genes from different breeds. These separate genetic effects all work together to form the breed group means shown in Table. These effects can differ greatly in size and in the direction that they operate. It isn't always easy to judge additive, maternal, and heterosis effects by looking at breed group means. Results for stillbirths were included in this analysis because the approach of examining additive, maternal, and heterosis effects showed differences between Holstein and Jersey gene sources.

The statistical procedure used estimates the "odds ratio" that Holstein genes in the calf (additive gene effects) would lead to more (>1.0) or less (<1.0) calving difficulty (or stillbirths) than Jersey genes. The design of the study allowed comparison of the odds that Holstein genes in the dam (maternal effects) would increase likelihood of difficulty giving birth (or to stillbirths) versus. Jersey genes in the dam. Finally, odds ratios were evaluated for the joint effects of pairs of genes (heterosis) either crossbred combinations (HJ or JH) versus the average of purebred combinations (HH or JJ). Table below shows the outcome of that analysis.

Holstein calves are 34.9 times as likely to experience dystocia at birth as Jersey calves. However, contrary to what some may believe, Holstein maternal genes are only 30% as likely to cause dystocia or stillbirths as Jersey maternal genes. However, negative additive effects of Holstein genes more than offset the maternal advantage.

Dairy farmers will certainly have fewer overall dystocia problems by adding Jersey genes to a crossbreeding program than by adding Holstein genes. There is no difference between crossbred gene combinations (heterosis) or the average purebred gene combinations for either dystocia or stillbirths when using Holsteins and Jerseys in the crossbreeding system. Holstein and Jersey breed differences in incidence of stillbirths are much less than differences in dystocia. These results are still preliminary, as about 25% of the births in this project are yet to be recorded, including a number of births of Jersey calves.

Table: *Risk (odds ratio) of dystocia or stillbirths from additive or maternal effects of Holstein genes vs. Jersey genes or heterosis (crossbred calves vs. purebred calves).*

Gene Effect	***Odds Ratio for Holstein vs. Jersey Genes***	
	Dystocia	***Stillbirths***
Additive	34.9	5.9
Maternal	0.3	0.3
Heterosis	1.4[2]	1.0[2]

Odds ratio greater than 1.0 indicates a greater probability of dystocia or stillbirths from Holstein genes than from Jersey genes.

Heterosis was not significant. Crossbred calves were equally likely to experience dystocia or stillbirths as purebred calves.

Table: *Comparisons of yields for Holsteins, Jerseys, and reciprocal crosses in the Virginia Tech-Kentucky crossbreeding project.*

Trait	***40 HH cows***	***27 HJ cows***	***23 JH cows***	***16 JJ cows***
Age at Calving, Months*	24.6	24.0	23.7	25.0
305d Actual Milk, lb	21579	18935**	20419	15244**
305d Actual Fat, lb	806	863	806	703**
305d Actual Protein, lb	645	643	643	500**
Peak Milk, lb	81	78	76	55**
Summit Milk, lb	74	68	70	53**

*Based on 122 cows that have freshened.

**Different from Holsteins ($P < 0.05$).

No statistical differences were found in age at first freshening between breed groups. All 122 cows that have freshened for the first time were used in this comparison, but 16 of these have not been in milk long enough to contribute to the other traits in the table. The table does not include any mature equivalent (ME) averages because there are no age adjustment factors for crossbreds. You will see ME values associated with crossbred records, but they are calculated from purebred age adjustment factors.

Thus far, HJ and JJ groups are producing significantly less than purebred Holsteins, but there is no significant difference between the JH and HH group for actual milk yield. Differences between Holsteins and crosses for fat and protein are not statistically different. Peaks and summit yields are numerically higher for Holsteins than crossbreds, but they are not significantly higher.

They are also significantly lower in production of milk, fat, and protein than the crossbreds. Unfortunately, the highest Net Merit Jersey bull has yet to contribute a daughter to the purebred Jersey group, but he has contributed to the JH group. The project will produce more data in coming months, and conclusions should wait for a more complete story.

3

Factors Influencing the Quantity and Quality of Milk

The Physiology of Mammary Glands During the Dry Period and the Relationship to Infection

The mammary gland of the dairy cow requires a non-lactating period prior to an impending parturition in order to optimize milk production in the subsequent lactation. This non-lactating period is commonly referred to as the "dry period" and is required for reasons other than nutrition. While dry periods can be of any length, there is general agreement that an approximate 60-day non-lactating period is required and that dry periods of less than 40-50 days result in less than optimal milk yields.

Mastitis research workers have observed for many years that the dry period is critically related to the dynamics of intramammary infection (IMI) within a dairy herd. New IMI during the dry period contributes in a major way to the increasing number of infected quarters that occurs with each successive lactation.

This critical relationship of the dry period to bovine mastitis resulted in the development of "dry cow therapy" as a necessary component of mastitis control. While it can be emphatically stated that dry cow therapy, when practiced in conjunction with post-milking teat disinfection, constitutes one of the most significant advances yet achieved in mastitis control, there is adequate reason to continue to investigate the relationship between intramammary infection and the dry period.

Firstly, as presently formulated and administered, dry cow therapy products are not equally effective against all pathogens. Secondly, the susceptibility of the mammary gland to new IMI by a particular pathogen may vary from one portion of the dry period to another, and thus, while new IMI by Streptococcus agalactiae or Staphylococcus aureus most likely occurs during the early dry period, new IMI by the environmental streptococci or coliform bacteria may occur frequently during the immediate prepartum period. A third reason for continued investigation is that the dry period, and in particular the early dry period, affords a unique opportunity to manipulate factors of natural resistance which are present in mammary secretion and which normally elevate markedly during the process of involution.

Physiological Function During the Dry Period

The physiology of the mammary gland during the dry period differs markedly from that during lactation. During lactation the primary function of the mammary gland is one of constant and continuous synthesis and secretion of milk. However, during the dry period the gland can be characterized as progressing through three distinct functional states, which are 1) the period of active involution; 2) the period of steady state involution; and 3) the period of lactogenesis and colostrogenesis.

The period of active involution begins with the cessation of regular milking and is probably completed by 30 days into the dry period. The period of steady state involution does not have definitive points by beginning and ending but represents the period of time the gland is maintained in the fully involuted state. The length of this functional state will increase or decrease proportional to the length of the dry period and is in contrast to the other two functional periods which are controlled by hormonal or management events. The third functional state is that of lactogenesis-colostrogenesis and is characterized by regeneration and differentiation of secretory epithelial cells, selective transport and accumulation of immunoglobulin and the onset of copious secretion. The active period of lactogenesis and colostrogenesis most likely begins 15-20 days prepartum.

Thus, at both the beginning and at the end of the dry period, the mammary gland is undergoing transition either from, or to, a state of active synthesis of copious quantities of secretion. The bulk of research evidence available would suggest that these periods of transition from one functional state to another, occurring in the absence of regular twice daily milking, are periods of heightened susceptibility to new IMI.

Active Involution-The Early Dry Period

The dry period is initiated by cessation of milking at the end of lactation. However, the mammary gland continues to synthesize and secrete milk which accumulates in the gland. Studies conducted with cows producing 9-10 kg of milk per day at drying off showed that mammary glands accumulated 75-80% of their daily yield at the time of drying off and that maximum accumulation of fluid occurred between 2 and 3 days post drying off. Fluid volume continues to decrease through at least day 16 of the dry period and probably through 30 days of involution.

The increase in intra-alveolar pressure as a result of milk accumulation is thought to trigger the events of active involution. An early event in involution is the appearance of lysosomes in the secretory epithelial cells. The lysosomes are thought to participate in autophagocytosis of secretory cell constituents. As the cells are degraded there is a loss of cell to cell contact as well as a loss of contact with the basement membrane. Heterophagocytosis of the degraded epithelial cells and the accumulated fat and casein is thought to be a function of the macrophages which enter the mammary tissue and secretion in large numbers during active involution.

As a result of these early events in involution, the composition of the secretion changed markedly. The rate of synthesis of the major milk constituents, i.e. fat, casein, lactose, citrate, â-lactoglobulin and á-lactalbumin decreases markedly by 3 to 4 days into involution. The actual concentration of these major constituents in the secretion of the involuting gland varies. The concentration of fat, casein, â-lactoglobulin and á-lactalbumin appears not to undergo major or consistent change. In contrast, the concentration of lactose and citrate, major regulators of osmolarity of mammary secretion, decrease markedly with involution but not prior to 3 days involution.

Proteins of mammary secretions whose origin is thought to be blood serum, are commonly present in whey of normal milk. The concentration of these proteins changes very little through 3 days involution, but by 7 days involution, the concentration of serum proteins in mammary secretion is significantly elevated. Thus, the marked increase in serum albumin and IgG immunoglobulin that occurs between 3 and 7 days if involution is consistent with the period of major reduction in mammary fluid volume. The concentration of all immunoglobulin classes, i.e. IgG_1,, IgG_2, IgA and IgM increases substantially by 7 days involution and there is a transitory increase in the selective transfer of IgG_1 which occurs at 2 to 3 days involution.

The magnitude of increase in selective transfer is far less than that observed during colostrogenesis, and the transitory nature of the increase may suggest that receptor sites required for selective transfer are either temporarily synthesized in greater numbers or made accessible as a result of the early involution events occurring in the secretory epithelial cell. The increase in serum albumin (BSA) probably reflects an increased permeability of the secretory epithelial cell barrier to passive diffusion of serum proteins. The concentration of BSA during involution never approaches the level found in blood serum or the concentration found in mammary secretion during acute inflammation. These findings suggest that the permeability barriers are not totally destroyed and that the mammary gland maintains a degree of control even though degenerative processes are occurring within the tissue.

A characteristic compositional change associated with involution is the dramatic increase in concentration of the iron biding protein lactoferrin (Lf). Lactoferrin is not specific to milk as it is present in the secretions of most mucosal epithelial surfaces throughout the body. Lactoferrin is a major constituent of the secondary granules of the polymorphonuclear neutrophil (PMN) and has been found in normal human blood serum at 0.8 ìg/ml. The major site of synthesis of the Lf found in bovine mammary secretions is thought to be the secretory epithelial cell with the PMN making only a minor contribution.

In normal bovine milk Lf is a minor whey protein and is found at 0.1-0.3 mg/ml. This is in marked contrast to the concentration in human milk (3-5 mg/ml) where it would appear to be the major whey protein. The concentration of Lf increases markedly in mammary secretion during active involution of the mammary gland in both the cow and the human. We have previously postulated that the increased numbers of lysosomes observed in secretory cells undergoing active involution is the source of the Lf observed in secretions of involuting mammary glands. The concentration of Lf is in the order of 0.4 mg/ml at drying off, increases to 1.4 mg/ml at 3 days involution, 8.5 mg/ml at 7 days involution and is generally found at 20-30 mg/ml by 30 days into the dry period. Thus, Lf is a major protein in the secretion of the non-lactating mammary gland.

Lactoferrin is thought to participate in the non-specific defence of the involuting mammary gland and has been reported to be of particular importance with regard to coliform infection. Lactoferrin is bacteriostatic by virtue of its ability to bind iron with great affinity. Thus, invading bacteria are forced to compete with Lf for iron which is required for normal growth of most pathogens. The possible exception is the streptococcal species which apparently have an extremely low iron requirement. However, there are also indications that Lf or peptides derived from LF can be bactericidal.

With regard to coliform infection, the bacteriostatic properties of Lf are reversed in the presence of citrate, a normal constituent of milk. Citrate, like Lf, is a chelator of iron and coliform bacteria that can utilize citrate-bound iron. The molar ratio of citrate to Lf determines the degree of coliform growth inhibition in vitro. As the molar ratio decreased, inhibition of coliform growth increased.

During involution the molar ratio of citrate to Lf decreases significantly. Examples of changes in the molar ratios are 2,356, 1,197, 581 and 49 at 0, 2, 3 and 7 days involution, respectively. The coliform growth inhibitory properties of whey collected during the first seven days of involution increased as the citrate to Lf molar ratio decreased. Thus, it would appear that the citrate to Lf molar ratio of bovine mammary secretions is inversely related to the ability of that secretion to inhibit coliform growth. In addition, the molar ratio of citrate to Lf serves as a sensitive marker system for the processes of involution and lactogenesis.

While there are numerous reports in the literature supporting the concept of Lf being a mediator of non-specific resistance to infection, there is a rapidly growing body of evidence suggesting that Lf may play an important role in the modulation and control of the functional processes of macrophages, lymphocytes and the PMN. Lactoferrin has been shown to interact specifically with receptors on the surface of the macrophage and the lymphocyte. As previously mentioned, Lf is a normal constituent of the PMN secondary granule and limited binding of Lf to the PMN surface has been reported. Lactoferrin may have an autoregulatory role to retain PMN at inflammatory sites and thus amplifies the inflammatory response.

Lactoferrin has been reported to be involved in the control of granulopoesis by inhibiting the production of granulocyte colony stimulating factor by the macrophage. There are reports that Lf can modulate lymphocyte function, and these reports suggest that Lf suppresses the response of these cells to antigen or mitogen. In addition, the iron binding proteins, namely Lf, transferrin and ferritin, play a role in directing the migration of cells with appropriate receptors and Lf may play a role in directing migration of mononuclear cells into acute inflammatory lesions and certain areas of lymphoid tissue.

That Lf receptors have been demonstrated on the surface of the macrophage and lymphocyte suggests that Lf may direct or control the influx of these cell types into mammary tissue during involution, and there is reason to speculate that once in the tissue of the involuting mammary gland, the high concentration of Lf may influence or modulate their functional activity.

The major cell types present in the secretion of the involuting bovine mammary gland are the macrophage, lymphocyte and PMN. Data consistently show that there are virtually no free secretory epithelial cells in the secretion of the involuted gland. The concentration of cells increases during the first week of involution, but prior to three days involution, there would appear to be very little increase over the level present at drying off. This is consistent with the lack of change in other parameters during the first three days of involution.

Mammary quarters infected at drying off or quarters that become infected have higher cell counts during involution than do uninfected quarters and the PMN would appear to be the predominant cell type present in the secretion during the first 7 days of involution. By 14 days involution, the macrophage is the dominant cell type present, but lymphocyte numbers are also increased. Lymphocyte numbers continue to increase between 14 and approximately 30 days involution, and this is in contrast to the marked decline in PMN numbers and a lesser decline in macrophage numbers over this period.

The lymphocyte would appear to be the major cell type present in mammary secretion at 30 days involution. The functional role of the phagocytic cells during early involution is generally thought to be that of phagocytosis of degenerated secretory epithelial cells, fat and casein. Presumably the macrophage plays the major role in removing the large quantities of accumulated fat and cellular debris. The marked decline in fluid volume and the reduction in the quantity of fat and casein between 3 and 7 days involution, would suggest that during this period virtually all phagocytic cells would be functionally committed to clearing the gland of these accumulated secretory products.

The functional role of lymphocytes present in the secretion of the involuting gland is virtually unknown. The lymphocytes would appear to consist of both B and T cell populations in ratios comparable to blood. A consistent finding has been that lymphocytes from bovine mammary glands are in general less responsive to mitogens than lymphocytes isolated from blood. There is some suggestion that many of these lymphocytes may have already undergone stimulation as a result of their presence in the secretion. On the other hand, Lf reduced the responsiveness of blood lymphocytes to the mitogen concanavalin A. The high concentration of lymphocytes in the involuting gland has been interpreted to suggest that the involuting or involuted gland would be more responsive to local immunization, and data have been published which support this hypothesis.

The Relationship of Active Involution to New IMI

There is little doubt that the rate of new IMI increases markedly during the process of active involution. The changes in mammary function and composition of the secretion of the gland could be interpreted to relate both positively and negatively to new IMI. Factors which may favour new infection during the early dry period are the following.

1) A large volume of milk is accumulated and leakage of milk from teats is often observed.
2) The contents of the gland are no longer being removed at regular intervals.

3) Teat end disinfection is stopped.
4) There is only minimal change in the composition of fluid through the first 3 days of involution. Thus, the factors of natural defence are only marginally elevated over normal milk levels during this potentially vulnerable period.
5) The phagocytic cells, a primary means of eliminating bacteria which gain entrance to the mammary gland, are occupied in the phagocytosis of degenerated epithelial cells, fat and casein.
6) There is reason to suspect that there are factors (other than fat and casein) present in the secretion of the involuting gland that may reduce the phagocytic capabilities of the macrophage and PMN and reduce the responsiveness of lymphocytes to antigen stimulation.
7) There would appear to be secretion components which are either increased in concentration or are produced during the process of involution, which can act as growth stimulants for certain bacteria.

From a positive point of view, the elevation in phagocytic cells, lymphocytes, immunoglobulins and Lf, could all be interpreted as changes occurring during involution that should render the gland more resistant to new IMI. However, it is likely that these changes occur too slowly and that their antibacterial action is compromised (as previously discussed) by other normal events of involution. Indeed, treatment of glands at or near drying off with colchicine, endotoxin or the combination of the two compounds resulted in a more rapid involution, earlier elevation of resistance factors and reduced IMI during the first week of the dry period.

Steady State Involution-Relationship to New IMI

Modern methods of dairy cattle management which stress minimum calving intervals and dry periods of 60 days or less, are likely to result in the process of involution (regression of the secretory tissue to terminal ducts with minimal numbers of alveoli present) being completed just prior to the hormonally initiated process of active lactogenesis and colostrogenesis. We would suggest that the overlapping of the two events may lead to a less than optimal hormonally mediated lactogenic response within the tissue and thus relate to the less than optimal milk yields in lactations following dry periods of under 40 days. As a consequence of dry periods of minimal length, the mammary glands of normal pregnant dairy cows may be maintained in steady state involution for a very short time period, and in reality, the gland would progress from active involution to active lactogenesis. However, in cows that are not pregnant or when dry periods are 60 days or greater mammary glands do enter into a period of steady state involution.

During this period fluid volume is maintained at levels of only a few ml, and the major constituents of milk are greatly reduced or at minimal concentration. Lactose and citrate are at concentrations of less than 5 mg/ml and.2 mg/ml, respectively. The citrate to Lf molar ratio is maintained 100 fold lower than that found during lactation. The concentration of immunoglobulins and Lf are elevated but generally at levels less than those recorded at 4 weeks involution. The predominant cell type would appear to be the lymphocyte in uninfected quarters but substantial numbers of macrophages are present. The PMN are generally found in low numbers by comparison to lymphocytes and macrophages.

The conditions within the mammary gland during steady state involution could be argued to favour the host in most relationships between host and pathogen. Indeed, the incidence of new IMI during this period has been reported to be the lowest of any functional state of the mammary gland. The degree to which the secretion of the fully involuted gland can resist the establishment of new IMI may vary with the bacterial species. The fully involuted gland is apparently quite resistant to the establishment of coliform infections following infusion of the organisms through the streak canal but susceptible to Str. uberis and Str. agalactiae.

Thus, the marked reduction in new IMI during steady state involution may relate partially to the high levels of antibacterial factors in the secretion but may also suggest a reduced rate of bacterial penetration through the streak canal. The latter is a result of a subumlike plug of keratin within the streak canal or a substantially lower level of exposure of teat ends to potential pathogens.

In support of the contention that the secretion of the fully involuted gland is highly antibacterial, it should be pointed out that spontaneous elimination of intramammary infection is a frequent event in the fully involuted gland and that acute clinical cases of mastitis are seldom seen in the fully involuted gland.

Lactogenesis-Colostrogenesis: Relationship to New IMI

The physiology of the mammary gland during the late dry period is again characterized by transition. In contrast to the early dry period the dramatic changes occurring in the mammary tissue are constructive as opposed to the destructive processes of active involution. In addition, the change are hormonally mediated. A major function of the mammary gland during this period is the formation and accumulation of colostrum and a unique feature of colostrum is the high concentration of the immunoglobulin IgG_1. The major physiological role of colostrum is that of a source of passively acquired antibody for the newborn calf.

The concentration of the major components of milk begin to increase in secretion approximately two weeks prior to parturition with marked increases in fat, casein and lactose occurring in the 5 days preceding parturition. The volume of fluid in the gland cistern increases slowly over the last two weeks of the dry period and then increases dramatically 1-3 days prepartum with the onset of copious secretion. The rapid increase in fluid volume is highly correlated with the dramatic increase in the concentration of citrate. The increasing concentration of citrate as a "harbinger" of lactogenesis or the onset of copious secretion.

During the period beginning approximately two weeks prepartum, the concentration of Lf declines to levels of 2-3 mg/ml found in colostrum. As a result of the changes in citrate and Lf, the molar ratio (citrate:Lf) increases by approximately 100 fold over the last few days prepartum. This dramatic increase in the citrate to Lf molar ratio suggests a loss of the antimicrobial properties of Lf and particularly in regard to coliform bacteria.

The concentration of immunoglobulin and in particular the IgG_1 begins to increase 2-3 weeks prepartum. Maximum concentration of all Ig classes is achieved approximately 5-10 days prepartum. The concentration falls as fluid accumulates in the gland with the onset of copious secretion. Approximate concentrations of IgG_1, IgG_2, IgA and IgM in colostrum are 60 mg/ml, 2.5 mg/ml, 4.5 mg/ml and 6.0 mg/ml, respectively.

Number of cells in secretion of uninfected quarters gradually decline over the 2 week prepartum period. The macrophage would again appear to be the predominant cell type. This is in contrast to the fully involuted gland where the lymphocyte is the most numerous cell. A most striking feature of the changing cellular content is the approximate 100 fold decline in lymphocyte numbers during the period 2 weeks prepartum. That the decline in lymphocyte numbers occurs coincidentally with the decrease in Lf concentration is of interest in light of the hypothesis that iron binding proteins may be involved in directing lymphocyte traffic into tissues.

The phagocytic ability of macrophages and PMN appear to be inhibited by the secretion of the prepartum gland or their function is reduced as a result of a general immunosuppression known to exist around parturition.. Macrophages isolated from colostrum are observed to contain numerous fat globules and may be one factor limiting phagocytic potential. However, research has suggested that inhibitors other than fat and casein may be present in the secretion of the prepartum mammary gland. The period immediately prepartum is known to be a period of increased susceptibility to new IMI caused by the environmental pathogens and dry cow therapy products as currently formulated and administered afford little or no protection to the gland during the latter portion of the dry period.

Management of Dairy Cows

Economic Characters in Dairy Cattle

The various economic characters in Dairy Cattle management are:

1. Lactation yield
2. Lactation period
3. Persistency of yield
4. Age at first calving
5. Service period
6. Dry period
7. Inter calving period
8. Reproductive efficiency
9. Efficiency of feed utilization
10. Disease resistance.

Lactation yield: The lactation yield in a lactation period is known as lactation yield. 'Thelactation yield in Indian breeds is very low compared to exotic breeds. This is dependent on no.. of calving, frequency of milking, persistency of yield (Normally in dairy cattle 30-40 % increase in milk production from first lactation to maturity is observed. After 3 or 4 lactation the production starts declining. For comparison of milk yield of different breeds and animals the mil yield should be converted into fat corrected milk (FCM).

4% FCM = 0.4 total milk + 15 total fat. After parturition the milk yield per day will be increased and reaches peak within 2-4 weeks after calving. This yield is known as peak yield. The maintenance of peak yield for more time is importance for better milk production. The lactation period in Indian breeds is low and so the production is also less and conversion.

Lactation period :The length of milk producing period after calving is known as lactation period. The optimum lactation period is 305 days. The milk production will Breeding of dairy animals and farm records 33 be less, if this period is shortened". Indian breeds will have less lactation period, but in some breeds this period is more with very little milk production Persistency of Milk Yield :During lactation period the animal reaches maximum milk yield per day with in 2-4 weeks which is called peak yield.

For high level of lactation yield, this peak yield should be maintained for longer period as far as possible, The maintenance of peak yield for long period is known as persistency, slow decrease in dairy milk yield after reaching peak yield in necessary. High persistency is necessary to maintain high level of milk production Age at first calving: The age o the animal at first calving is very important for high life time production. The desirable age at first calving in Indian breeds is 3 years, 2 years in cross breed cattle and 3 1/2 years in Buffaloes. Prolonged age at first calving will have high production in the first lactation) but the life time production will be decreased due to less no of calvings. If the age at first calving is below optimum, the calves born are weak, difficulty in calving and less milk production in first lactation.

Service period : It is the period between-date of calving and date of successful conception. The optimum service period helps the animal to recover from the stress of calving and also to get back the reproductive organs back to normal For cattle the optimum service period is 60-90 days. If the service period is too prolonged the calving interval prolonged, less no. of calvings will be obtained in her lifetime and ultimately less fife time production.' If the service period is too short, the animal will become weak and persistency of milk production is poor due to immediate pregnancy.

Dry Period: It is the period from the date of drying (stop of milk production) to next calving When the animal is in pregnancy, before next calving. The animal should be given rest _period to compensate for growth of foctus. A minimum of 2-234 Dairy Animal Management½ months dry period should be allowed) If the dry period is not given or too low dry period, the animals suffer from stress and in next lactation, the milk production drops substantially and also it gives weak calves. On the other hand if the dry period given is too high, it may not have that much effect on increasing milk yield in the next lactation, but it decrease the production in the present lactation.

Inter calving period : This is the-period between two successive calvings. It is more, profitable to have one calf yearly in cattle and at least one calf for every 15 months in buffaloes. If the calving interval is more, the total no. of carvings in he life time will be decreased and also total life production of milk. decrease. Reproductive Efficiency : The reproductive efficiency means the more number of calves during lifetime, so that total life time production is increased, The reproduction or breeding efficiency is determined by the combined effect of hereditary and environment. Several measures of breeding efficiency like number of services per conception, calving interval, days from first breeding to conception are useful.) Reproductive efficiency has generally a low heritability value indicating that most of the variations in this trait is due to non genetic factors. In adverse environmental conditions, the poor milk producing animals may not be much effected compared to high effect in high milk yield.

Efficiency of Feed Utilization and Conversion into Milk : The animal should utilize the feed efficiently to convert into the milk. Disease Resistance : Indian breeds are more resistant to majority of disease compared to exotic cattle. Cross breeding helps to get this character.

System of Breeding

Breeding is defined as the crossing of the male and the female parents to get the off spring for the characters desired.

Housing of Dairy Cattle

Selection of Site for Dairy Farm: The points which should be considered before the erection of dairy buildings are as follows :

Topography : A dairy building should be at a higher elevation than the surrounding ground to offer a good slope for rainfall and drainage for the wastes of the dairy to avoids tagnation within. A levelled area requires less site preparation and thus lesser cost of building. Low lands and depression and proximity to places of badod our should be avoided.

Soil Type: Fertile soil should be spared for cultivation. Foundation soil as. far as possible should not-too dehydrated or desiccated. Such a soil is susceptible to considerable swelling during rainy season and exhibit numerous cracks and fissures.

Exposure to the sun and protection from wind : A dairy building should be located to a maximum exposure to the sun in the north and minimum exposure to the sun in the south and protection from prevailing strong wind currents whether hot or cold. Buildings should be placed so that direct sunlight can reach the platforms, gutter and mangers; in the cattle shed. As far as possible, the long axis of the dairy barns should-be set in north south direction to have the maximum benefit of the sun.

Accessibility: Easy accessibility to the buildings is always desirable. Situation of a cattle shed by the side of the main road preferably a distance of about 100 meters should be aimed at.

Durability and Attractiveness : It is always attractive when the buildings open upto a scenic view and add to the gradeur of the scenery. Along with this, durability of the structure is obviously an important criteria in building a dairy.

Water Supply : Abundant supply of fresh, clean and soft water should be available at a cheap rate.

Surrounding : Areas infested with wild animals and 'daoits; should be avoided. Narrow gates, high manger curbs, loose hinges, protruding nails, smooth finished floor in the areas where the cows move and other such hazards-should be eliminated.

Labour : Honest, economic and regular supply of labour be available.

Marketing : Dairy buildings should only be in those areas from where the owner can sell his products profitability and regularly. He should be in a position to satisfy the needs of the farm within no time and at a reasonable price.

Electricity : Electricity is the most important sanitary method of lighting a dairy. Since a modern dairy always handles electric equipments which are also economical, it is desirable to have an adequate supply of electricity.

Facilities, labour and food : Cattle yards should be so constructed and situated in relation to feed storages, hay stacks, silo and manure pits as to effect the most efficient utilization of labour. Sufficient space per cow well arranged feeding mangers and resting areas contribute not only to greater milk yield of cows and make the work of the operator easier but also minimise feed expenses. The relative position of the feed stores should be quite adjacent to the cattle barn. Noteworthy features of feed stores are given below.

a) Feed storages should be located at hand near the centre of the cow barn.

b) Milk house should be located almost at the centre of the barn.

c) Centre cross-alley should be well designed with references to feed storage, the 'stall areas and the milk house.

Systems of Housing-Loose Housing System

Conventional Dairy Barn

The most widely prevalent practice in this country is to tie the cows with rope on a Katcha floor except some organised dairy farms belonging to government, co-operatives or military where proper housing facilities; exist. It is quite easy to understand that unless cattle are provided wit good housing facilities, the animals will move too far in or out of the standing space, defeating all round and even causing trampling an wasting of feed by stepping into the managers. The animals will be exposed to extreme weather conditions all leading to bad health and lower production.

Dairy cattle may be successfully housed a wide variety of condition, ranging from close confinement to little restrictions except at milking time.

1. The loose housing barn in combination with some type of milking barn or parlour.
2. The conventional diary barn.

Loose Housing System

Loose housing may be defined as a system where animals are kept loose except milking and at the time of treatment. The system is most economical. Some features of loose housing system are as follows.

1. Cost of construction is significantly lower than conventional type.
2. It is possible to make further expansion without change
3. Facilitate easy detection of animal in heat.

4. Animals feel free and therefore, proves more profitable with even minimum grazing
5. Animals get optimum excise which is extremely important for better health production.
6. Over all better management can be rendered.

Sanitation in Dairy Farm

Sanitation is necessary in the dairy farm houses for eliminations of allmicro organism that are capable of causing disease in the animals. The presence of organisms in the animal shed contaminates the milk produced thus reducing its self life, milk produced in an unclean environment is likely to transmit diseases which affect human health: Dry floorings keeps the houses dry and protects from foot injury. Similarly the presence of flies and other insects in the dairy farm area are not only, disturbs the animals but also spreads deadly diseases to the animals egg. Babesiosis, Theileriosis.

Cleaning Of Animal Sheds: The easy and quick method of cleaning animal house is with liberal use of tap water, proper lifting and disposes all of dung and used straw bedding, providing drainage, to the animal house for complete removal of liquid waste and urine. The daily removal of feed and fodder left over in the manger, reduces the fly nuisance. Periodical cleaning of water through eliminates the growth of algae, bacterial and viral contamination and thus keeps the animal healthy.

Sanitizers : Sunlight is the most potent and powerful sanitizer which destroy most of the disease producing organism. Disinfection of animal sheds means making these free from disease producing bacteria and is mainly-carried out by sprinkling chemical agents such as bleaching powder, Iodine and lodoph or, sodium carbonate, Washing soda, Slaked Lime (Calcium hydroxide), Quick Lime (Calcium oxide) and phenol.

Bleaching Powder: This is also called calcium hypo chloride. It contains upto 39 % available chlorine which has high disinfecting activity.

Iodine and lodophor :-This is commercially available as lodophores and contains between 1 and 2 % available Iodine which is an effective germicide.

Sodium Carbonate: A hot 4 % solution of washing soda is a powerful disinfectant against many viruses and certain bacteria.

Slaked Lime And Quick Lime : White washing with these agents makes the walls of the sheds and the water troughs free from bacteria.

Phenol : Phenol or carbolic acid is a very disinfectant which destroy bacteria as well as fungus.

Insecticide : Insecticides are the substances or preparations used for killing insects. In order to control flies and disease transmitting ticks, insecticides are used in dairy farms. Ticks usually hide in cracks and crevices of the walls and mangers. Smaller quantities of insecticide solutions are required for spraying. Liquid insecticides can be applied with a powerful sprayer_ hand sprayer, a sponge or brush, commonly used insecticides are BHC, DDT, Gramaxane wettable powders, malathion, surriithion, Sevin 50 % emusifying concentration solutions.

These are highly poisonous and need to be handled carefully and should not come in contact with food material, drinking, water, milk etc.

Precautions While using disinfection I Insecticide:

- Remove dung and used bedding completely.
- Avoid spilling of dung and used bedding while carrying it out.
- Avoid the use of dirty water in cleaning the sheds.
- Never put the fresh fodder over: the previous day's left over fodder in the manger.
- Prevent algae to grow in the water troughs
- Use proper concentration of disinfectant/insecticide solutions to avoid any toxic effects poisoning.
- Avoid of the mat the milking time as milk absorbs these quickly.

Procedure:

- Remove the dung from the floor and urine channel with the help of a shovel and basket (iron) and transfer it to the wheel-barrow. Remove the used bedding and leftovers from the mangers in a similar way.
- Empty the water trough and scrape its sides and bottom with the help of a floor brush.
- Wash the water trough with clean water and white wash it with the help of lime mixture once a week.
- Scrape the floor with a brush and broom and wash with water.
- Clean and disinfect the splashes of dung on the side walls, railing and stanchions.
- Remove the cobwebs periodically with the help of a wall brush.
- Sprinkly one of the available disinfecting agent in the following concentration. Bleaching powder should have more than 30% available chlorine. Phenol 1-2% solution. Washing Soda (4% solution).
- Allow adequate sunlight to enter in to the shed.

- Spray insecticides at regular intervals specially during the rainy season (Fly season).
- White wash the walls periodically by mixing insecticides in it to eliminate ticks and mites living in cracks and crevices.

Care and Management of Calf

We must give good! feeding and management for the calves so that they develop well and,; useful for replacement stock. The feeding and care of the calf begins~ before its birth The dam should be dired 6-8 weeks before expected calving and should be fed well. Underfed animals will give weak and smog calves.

Early Management:

1) Immediately after birth remove any mucous or phlegm from those nose and mouth.
2) Normally the cow licks the calf immediately the birth. This helps' dry off the calf and helps in stimulating breathing and circulation. When the cows does not lick or in cold climate, rub and dry the calf with a dry cloth or gunny bag. Provide artificial respiration by compression and relaxing the chest with hands
3) The Naval should be tied about 2-5 cms away from the body and cut 1cm below the ligature and apply Tr. Iodine or boric acid or any antibiotic.
4) Remove the wet bedding from the pen and keep the stall very clean and dry in condition.
5) The weight of the calf should be recorded.
6) Wash the cow's udder and teats preferably with chlorine solution and dry. Allow the calf to suckle the first milk of the mother i.e. Colostrum.
7) The calf will be standing and attempts to nurse within one hour. Otherwise help too weak calves.

Feeding of Calves :

1) Feed colostrum i.e. the first milk of the cow for the first 3 days. The colostrum is thick and viscous. It contains higher proportions of Vit A and proteins. The proteins are immune globulin which gives protection against many diseases. colostrums contains antitrypsin which avoid digestion of immunoglobulins in the stomach and is absorbed as it is.
2) Whole milk should be given after 3 days it is better to teach to, drink the milk from the pail or bucket. Feed twice a day which should be warmed to body temperature. For weak calves feed thrice a day.

3) The limit of liquid milk feeding is 10 % of it's body weight with a maximum of 5-6 litres per day and continue liquid milk feeding for 6.10 weeks. Over feeding causes 'Calf Scours'.
4) The milk replaces can be given to replace whole milk.
5) Give calf starter after one month of age.
6) Provide good quality green fodder and hay from 4th month afterwards.
7) Feeding of antibiotics to calves improves appetite, increases growth rate and prevents calf scours. E.g. aureomycin, Terramycin etc.

Other Management Practices:

1) Identity the calf by tattooing in the ear at birth, and branding after one year.
2) Dehorn the calf within 7-10 days after birth with red hot Iron or caustic potash stick or electrical method.
3) Deworm the calf regularly to remove worms using deworming drugs. Deworm at 30 days interval.
4) Fresh water should be given from 2nd week onwards.
5) House the calves in individual calf pens for 3 months afterwards in groups. After six months males and females calves should be housed separately.
6) Weigh the calves at weekly interval upto 6 months arid at monthly interval afterwards to know the growth rate.
7. Mortality in calves is more in first month due to pnuemonia. Diarrhea (calf scous) and worms. House them under warm condition, clean condition to avoid above condition.
8. Extra teats beyond 4 should be removed at 1-2 months of age.
9. 8-9 weeks of age, males should be castrated.
10. Keep the body clean and dry to avoid fungal infection.
11. Mineral-blocks should be provided, so that the calves lick and no changes for mineral deficiency.
12. Wean the calf from the mother and feed through pail feeding system.

Care and Management of Heifer

Better Care and Management of heifer will give high quality replacement stock to the dairy farm. The following care and Management practices are recommended for a heifer.

1. Feed the heifer sufficiently to produce normal growth. During the early stage relatively more protein than energy is needed. Most heifers grow

well if excellent hay is given as much they can eat. The amount of growth depends upon the quality of forage fed.

2. The heifers should be provided with a dry shelter free from drafts. A loose housing system with a shelter open to one side is sufficient.
3. The size rather than the age of a dairy heifer at breeding time is important. Breeding under sized animals is never profitable. They may be stunted or slow to reach maximum size. Small heifer are more likely to have difficulty in calving. Though the heifer that is bred to calve at an older age yields higher milk yield in the first lactation, the total milk produced by such a cow will be less when compared to the heifers that freshens at an early age. Usually the heifer is bred to freshen at 24-30 months of age.
4. The heifer should be growing and in good flesh at calving time. This is necessary so that she can produce milk at the most profitable level.
5. Place the heifer in a separate shed about 6-8 weeks before she is due to calve.
6. Feed 2-3 kgs of concentrate daily and all the forage she eats.
7. Before calving let the heifer becomes accustomed to handling and to the procedures used in the milking herd. Always handle her gently and with kindness.
8. Maintenance of health among heifers is very important for proper growth. The health among the heifers is maintained by hygienie housing, water balanced feeding and taking necessary preventive steps against common diseases.
9. Periodically the heifers in the herd should be checked for their proper growth and other progress. Animals lagging behind below the required standards should be removed from the herd.
10. For the heifer the calving is first time and it may have difficulty in calving. So take extra care during calving.

Care and Management of Milch Animal

To get high milk during any lactation, the milch animal should be properly fed and necessary care and managemental practices should be followed.

1. Provide green succulent forage together with leguminous hay or straw to the extent of animal can consume, so that all its maintenance requirements are met with through forage only. Extra concentrate at the rate of 1 kg for every 2 to 2.5 litres of milk should be provided. Salt and mineral supplements should be given to maintain the lactation.
2. Never frighten or excite the animals. Always treat them gently and with kindness.

3. With proper feeding and care, a cow will come to heat with in 16 days of calving. Do not with hold service unnecessarily after the signs of heat are noticed in a cow. The shorter the interval between calvings, the more efficient the animal is as a milk producer. By maintaining proper records of breeding and calvings of the animals will ensure a study flow of milk through out the year.
4. Individual attention to feed each animal according to its production is a must. For this purpose maintain individual production records.
5. Keep up regularity of feeding. Concentrate mix is fed before or during milking, when as roughages after milking. This practice will avoid dust in the shed.
6. Water should be provided to drink at will or at frequent intervals. It is more beneficial, if the animal is maintained on paddy straw as sole rough age..
7. Regularity in milking is essential. Increase of milk in the udder will reduce further secretion of milk. Milking thrice is better than twice since 10-15 % more milk can be produced.
8. Rapid, continuous, dry hand milking should be practiced without undue jerking of teats, milking should be done with whole hand, but not with thumb and index finger.
9. Cows should be trained to let down milk without calf suckling. This will held to wean the calves early.
10. Loose housing with shelter during hot part of the day should be provided. The animals will get maximum exercise in loose housing system.
11. Grooming of the cows and washing of the buffaloes before milking help in clean milk production. Daily brushing will remove loose hair an dirt from the coat. Grooming will also keep the animal hide pliable.
12. Wallowing of buffaloes or water spraying on their bodies will keel the buffaloes comfortable especially in summer.
13. Common ailments should be properly detected and treated.
14. Common vices should be properly detected and care should be taken. Eg. Kicking, licking, suckling etc.
15. Provide at least 60-90 days dry period between calvings. If the dry period is not sufficient, the milk yield is subsequent lactation will be reduced.
16. Vaccinate the cows-against important diseases and also guard against insects and pests.

17. Every animal should be numbered and particulars pertaining to milk, fat %, feed taken, breeding, drying and calving dates should be recorded.

Care And Management of Dry and Pregnant Animal

The good care and managemental practices given to pregnant animal will give good calf and also high milk yield during the successive lactation.

1. Extra concentrate mix of 1.25 to 1.75 kgs should be provided for pregnant animal as pregnancy allowance. Feed good quality of leguminous fodder. The animal should not be not-lean-not fat' condition.
2. Provide clean drinking water and protection from thermal stress.
3. Do not allow them to mix with other animals that have aborted or that are suffering from or carriers of diseases like brucellosit.
4. Allow moderate exercise, which helps in calving normally. Do not tire them by making long distances especially on uneven surfaces.
5. Do not allow them to fight with other animals and take care that they are not chased by dogs and other animals.
6. Avoid slippery conditions, which causes the animal to fall receiving fractures, dislocation etc.
7. If accurate breeding records are available, calculate the expected date of calving. Separate it one or 2 weeks before and shifted to individual parturition pens. These pens are thoroughly cleaned and fresh bedding may be provided.
8. Feed one kg extra concentrate during last 8 weeks of gestation. Feed laxative about 3-5 days before and after calving (Wheat bran 3 kgs + 0.5 gs of Groundnut cake + 100 gms of mineral mixture of salt).
9. Symptoms of delivery may be observed i.e. swelling of external genetalia, swelling of udder, usually majority of animals will deliver without any help. If there is any difficulty, provide veterinary help.
10. After parturition external genetalia, flank should be cleaned arid protect the animal from chill and give warm wafer.
11. Placenta will normally leave the cow within 2-4 hours after calving. It not take the help of a veterinarian.
12. Take care of the animal before calving from milk fever. Give calcium supplement.
13. Some times the udder will be swollen just before calving. Remove the milk partially.
14. Take care, of the animal, if at all any abortion.
15. Provide always free access to drinking water.

Care and Management of Bulls and Bullocks

Care and Management of Bull

The maintenance of breeding bulls in good condition and suitable for breeding is highly essential requirement for the success of breeding programmes. A rising condition is better for reproduction than a falling one. Fat males may produce semen of inferior quality or they may be slow or fail at service. Breeding bull should receive plenty of exercise, will usually produce large ejaculation containing more sperms of higher activity.

A breeding bull should housed separately known as "Bull Shed" with sufficient area of floor and proper covering. It is sound practice to provide cool conditions and adequate drinking water. A balanced rations should be fed containing adequate energy, proteins, minerals, and vitamins. Green fodder must be available both before and during breeding season.

Most of the bulls are ferocious and so control them properly using nose rings etc. It is of great importance that males should be, fed regularly and not too much at one time, and too little at another. For bulls two mating a day has been found to be openings.

Moderate exercise should be provided to keep the breeding bull in active and non fatty conditions. Regular grooming of the breeding bull be practiced. In buffalo bulls regular shaving may be practiced.

Care and Management of Bullock

Bullocks are normally used for agricultural operations and or transport purpose. Some bullocks are ferocidus and so control them properly with nose rope or nose rings. The hooves of the bullocks should be provided with metal shoes to protect the hooves from wear and tear. The working hours for bullocks are recommended as follows :

a) Normal Work-6 hours of carting or 4 hours of ploughing.

b) Heavy Work-8 hours of carting or 6 hours of ploughing.

Sufficient roughages and 1-2 kgs of concentrates may be provided for feeding of bullocks during break period in works, the animal may be left for free grazing.

The bullocks are housed in separate sheds with sufficient space and protection from hot and cool conditions. Free access to drinking water is essential. Regular grooming of animals should be practiced.

Management of the Dairy Cow

There are thousands of dairy farms around the world with different farm plans and management strategies. Climate, market conditions, traditions, breeds etc. affect the running and planning of the business. If we try to categorise and generalise the world wide dairy operations, we relatively easily find three main "types". The American type (North and South America) is characterised by large loose-housing operations, total mixed ration feeding (TMR) and relatively many employees.

However, dairy farms in Northeast US and parts of Canada differ from the typical American operation. There you find many smaller family farms with either loose-housing or stanchion barns. These operations are quite similar to the European type, which is characterised by relatively small operations where each cow is fed and treated individually.

The third type is mainly found in New Zealand and Australia. The dairy operations in this area are usually run in a very extensive way, relying to a great extent on grazing. Due to the climate and local restrictions, New Zealand and Australian dairy farmers do not rely on concentrate use so much. On top of that, the milk market is deregulated and the prices follow the world market price (WMP).

When you look closely at all these farm types, they share many underlying principles. Regardless of management strategy, farm size or local climate, the cow has to calf, produce milk, eat, be kept in good health status etc. The goal for any dairy farm should be to maximise profit per unit of milk within farm constraints and, in some parts of the world, within milk quotas. To be able to do so, we need to know how to manage the cow and how different production aspects interact with each other.

Feeding

As the dairy cows genetic potential is increasing, feeds and feeding strategies are becoming more and more important. It is well known that the amount of milk to be produced is highly influenced by the amount and quality of the feed given to the cow.

It is also possible to influence the milk composition through the feeding. As the cow normally experiences a shortage of nutrients in early lactation, it is of importance to feed the cow a well balanced diet in order to maximise the dry matter intake. An unbalanced diet increases the risk for metabolic disturbances and weight loss, which have a negative effect on the milk yield. Healthy cows will also make the transition from dry to peak easier.

The cow's largest compartment is the rumen. Together with the reticulum, it has a total volume of approx. 150.200 litres. In these two compartments, there are billions of micro-organisms. They help the cow to digest and utilise the nutrients in the feed. To reach a good feed utilisation, and in the end a high milk yield, the micro-organisms have to have optimal conditions.

Managing Feeding

One of the primary keys to a successful dairy operation is a good nutrition program. Not only is nutrition one of the highest input costs (about 50% of the total costs), but it also controls the results of milk production, reproduction and health. The basis in all feeding programs is the roughage and a common recommendation is that this shall account for 40.60 percent of the total dry matter intake.

A cows intake per day is limited, and it is therefore important to know the dry matter and nutritional content of both the roughage and concentrate. However, the cows daily dry matter intake is dependent on her stage of lactation. The intake will gradually increase after calving, to reach its peak approx. 6.12 weeks after calving.

When all the feedstuffs (including roughage) have been analysed, we have to make sure that each cow gets what she need in terms of energy, protein, minerals, vitamins and water. To do so there are a number of feeding strategies and systems to choose between. Their common trait is that the roughage intake is controlled through the distribution of concentrates. The more concentrates that are supplied, either separately or in a mix with the roughage, the less roughage the cow will eat.

Feeding Strategies

Ad Libitum Feeding

Ad libitum feeding simply means that the animals have free access to the feed and are allowed to eat as much as they want. The roughage is often fed ad libitum regardless of the strategy used for feeding the concentrates (flat rate feeding, challenge feeding and feeding to yield).

However, by mixing roughage and concentrates and feeding it ad libitum, the total feed intake for a group of cows can be regulated by means of the degree of concentration of the feed. This strategy is common when using a mixer wagon. A disadvantage with ad libitum feeding is that it is not possible to control each cows individual feed intake, which increases the risk of both over-and underfeeding.

Flat Rate Feeding

Flat rate feeding is a strategy in which all cows are fed the same quantity of concentrates during the entire (or part of) lactation. The concentrate is restricted on a certain level, while the roughage is fed either ad libitum or in a mix. As described earlier, the cows energy and nutrient demand varies depending on the stage of lactation. Because of the fixed concentrate ratio, flat rate relies on fat mobilisation. The disadvantages with this strategy are the risk for metabolic disturbances (Ketosis) and the difficulty to reach high peak yields. However, the technique is simple and therefore the investment cost can be relatively low.

Challenge Feeding

While flat rate feeding relies on fat mobilisation, challenge feeding and feeding to yield aims to supply the cow with nutrients that are needed for the actual lactation stage.

Challenge feeding of concentrates takes place in the early lactation, when there is a risk of underfeeding. Every cow is given the maximum ration of concentrates that she can consume, without reducing her roughage intake below approx. 40.60 % of total dry matter intake. This is continued until she reaches peak, four to ten weeks after calving. By using this strategy, each individual cow is given the possibility to show her production potential.

Feeding to Yield

When the peak production is reached and the production starts to decline, there is a risk of overfeeding. By feeding the cow to yield, which means that each cow is fed concentrates individually according to her actual milk yield, the cows body condition is maintained and the feed is used as efficiently as possible. A successful implementation of this strategy means increased milk production with the feed cost well under control.

Feeding Strategy

Ad libitum:

- High risk of over/ underfeeding
- Waste of feed
- Difficult to reach high peak production

Flat rate feeding:

- Simple ration
- Risk of overweight cows
- Risk of underfed cows which may cause Ketosis
- Difficult to reach high peak production.

Feeding Systems

The layout of the feeding system and the equipment involved differs depending on the type of barn, stanchion or loose-housing, and the management strategy used on the farm. However, the principles of the feeding systems are generally the same and can be divided into the following four categories: Manual Concentrate Feeding, Computerised Concentrate Feeding, Total Mixed Ration (TMR) and Partly Mixed Ration (PMR).

Manual Concentrate Feeding

This system is common in stanchion barns but it is also found in loose-housing barns using manually operated in-parlour feeding. In a stanchion barn, the system makes individual feeding of both roughage and concentrate possible. However, in both loose-housing and stanchion barns, the roughage is usually fed ad libitum and the concentrate is fed individually.

Suitable feeding strategies depend on the barn layout. In a stanchion barn it is possible, with a varying grade of precision, to implement challenge feeding and feeding to yield strategies. A loose-housing barn with manual in-parlour feeding is more or less only suitable for flat rate feeding.

A drawback with manual systems is the risk of low accuracy. Practical experience has shown that there often are great differences between the intended ration and actual ration. The diagram below shows that, for example, if an animal should have 4.5 kg, the actual ration could range from 3 kg to 8 kg. This may not only result in costly feed waste and overweight cows, but will also have a negative effect on the milk production. A manual system is also labour demanding and requires more time to implement new feed rations, compared to a computerised system. All this means high feed and labour costs and lost milk revenue opportunities.

Computerised Concentrate Feeding

Instead of feeding the concentrates manually, they can be dispensed and controlled by a computer. In stanchion barns, a rail suspended feed wagon is normally used.

For loose-housing systems, there are two systems to choose from: in-parlour feeding or out-of-parlour feeding (or a combination of both).

The in-parlour feeding is simply a dispenser that can be operated by manual or automatic identification. Out-of-parlour feeders requires automatic identification (transponders). Farms with a long grazing period sometimes prefer to feed all concentrates in the parlour. A disadvantage with this is that large rations cause long eating times and are unhealthy for the rumen. An alternative to this is to use out-of-parlour feeders all year round and keep the cows housed for a couple of hours before letting them out on pasture again.

When out-of-parlour feeding or a rail suspended feed wagon (stanchion barns) is used, challenge feeding and feeding to yield can successfully be used as the feeding strategies. Each cows individual ration is programmed in a computer and when the cow enters the feed station, she is identified and the correct amount of feed is dispensed. The computer decreases the demand for labour and facilitates the implementation of new feed rations and the record keeping for each cow.

Total Mixed Ration (TMR)

In the Total Mixed Ration feeding system, the concentrates and roughage are mixed in a mixer wagon and usually fed ad libitum. The mix is often dispensed to the cow directly from the mixer wagon, but can also be distributed by band conveyors or with rail suspended feed wagons. The TMR system is most common in large loose-housing herds, but can also be used in stanchion barns.

There are mainly two ways to feed the cows with TMR: either feed the same ration to the whole herd (no grouping) or different rations to different groups. In non-grouping systems, only one mix is made and dispensed.

The mix is usually composed to suit the high yielding cows, which makes the balance of the ration very important to avoid overweight cows. A way to improve the precision of the feeding is to use a grouping system. The cows can be grouped according to many criteria: yield, stage of lactation, first calf heifers etc. The drawback with grouping is that much of the rationality of the system is lost compared to the non-grouping system.

Much time must be spent moving the cows from one group to another as the production changes. Group changes also result in milk loss due to social adjustment. In addition, several mixes have to be prepared and dispensed to suit each group.

However, if the herd is large enough, it should be possible to run an efficient and convenient TMR system with grouping. Disadvantages with all TMR systems are the relatively high amounts of feed waste and the fact that it is not possible to monitor the individual consumption of concentrates. The cost of feeding and storage equipment is also relatively high.

The main advantage is good rumen health, which results in few metabolic disturbances and possible effects on the milk production. To obtain a well working TMR system and a high dry-matter intake, the ration requires much attention.

Partly Mixed Ration (PMR)

Partly Mixed Ration (PMR) is a feeding regime that combines TMR and individual feeding of concentrates. A mixer wagon is used to mix roughage and some of the concentrates. The concentrate level in the mix is adjusted to fit the lower yielding cows.

The high yielding cows are then fed extra concentrates by out-of-parlour feeders, in-parlour feeders or from a rail suspended feed wagon. The PMR system gives you the possibility to combine the advantages of a mixed ration with computerised feeding.

Milking

To illustrate an individual cows milk production, we normally plot the yields against time, which gives us the lactation curve illustrated in the picture below. As the diagram shows, the milk yield will rise during the first months after calving, which then is followed by a long period of continuous decline. The shape of the lactation curve will differ from individual to individual and from breed to breed. Feeding and management will also influence the shape and have a significant impact on the total amount of milk produced. Lactation is ideally 305 days, but in practice it is usually more, followed by a two-month dry period prior to the next calving. A cows milk yield is influenced by many factors, which are described more in detail in the DeLaval booklet Efficient Milking.

Peak Yield

Peak yield is the point where the cow reaches the highest milk production level during the entire lactation. Heifers peak at 70.75% of mature cows and second lactation cows peak at 90% of mature cows. Normally the peak is reached four to ten weeks after calving. The time it takes to reach peak yield varies with many factors, for example breed, nutrition and yield potential. Higher producing animals tend to peak later than low producing ones. A high peak yield normally means a higher total yield. Research shows that each one kilogram increase in peak yield usually means an additional 100.200 kg of milk produced during the actual lactation. Reaching high peak yields requires a very well managed and balanced feeding programme.

Persistency

After the peak, milk production starts to decline by approximately 7.10% per month. The rate of decline is normally measured as the cows persistency. If a cows milk production falls by 7% per month after peak, it is more persistent than a cow that loses 10% per month. A general rule of thumb is that higher peak production leads to lower persistency. As for peak yield, the persistency is dependent on feeding and is therefore, to some extent, possible to influence. Persistency differs between cows, but a first lactating cow is normally more persistent than a second or third lactating cow.

Total and Daily Milk Yield

Close monitoring and evaluation of each cow or group are important ingredients when trying to reach a high level of total milk production (high peak and long persistency). However, it is important to remember that a dairy cow is not a stable milk producer. The milk yield differs from day to day and the relative variation can be as high as 6.8 % from one day to another. Cows milked three times or more a day usually have a lower variation than cows milked twice a day.

Managing Milk Yield Recording

The total milk yield is a good indicator of gross income from milk sales, which has a direct effect on the dairy farms income. With milk yield recording, the dairy farmer can monitor his day to day output and use it to control and monitor the production. Milk yields are also the most interesting variable when calculating the feed rations. Common methods for measuring milk yield are: bulk tank monitoring, monthly milk yield recording, recorder jars/stand alone meters and electronic milk yield registration. The milk yield records can then be used for:

- Calculating and follow-up of the nutrient intake
- Monitoring long-term production changes
- Indicating overall health disturbances.

Bulk Tank Monitoring

Bulk tank monitoring does not enable you to monitor the cows individually, but it can be used as an indicator of the overall herd management. To get the correct total yield, you have to account for the number of fresh, treated and dry cows plus the milk used for own consumption and calves.

Monthly Milk Yield Recording

Monthly recording is usually carried out by a milk yield recording organisation and was originally introduced for breeding purposes. Due to the fact that the recording normally takes place once a month, and each cows milk production varies day to day, it is not recommended to use this information to manage feeding. When receiving the data, it will be historical and not reflect the current situation.

However, the information can be used for trend analysis and give an idea of the peak and persistency for individual cows, which can be used for culling, breeding and general feeding decisions. As for bulk tank monitoring, the monthly milk yield recording does not give the dairy farmer an early indication of health problems.

Daily Milk Yield Recording

Daily milk yield recording can be considered as one of the most important decision aids for fine tuning the high producing herd. The milk yield can be measured at every milking by using either recorder jars or milk meters. If using recorder jars, all data has to be recorded manually (e.g. on paper or typed into a computer).

With milk meters connected to a processor or computer, you will have an automatic collection of data and a memory capacity that is much more accurate and efficient than the milkers. The processor contains a database where all individual milk yields and other relevant data are stored on a daily basis. This data provides the manager with exact and timely information of a cows or groups production on a day to day basis, which can be used to:

- Calculate and evaluate feed rations for individual cows or groups to maximise the milk production and avoid over and/or under feeding.
- Identify cows in heat. Research shows that decline in morning milk might indicate oestrus.
- Identify cows with potential health problems early. For example, ketosis is related to a gradual decrease in milk production before it can be diagnosed.
- Calculate the actual lactation curve instead of estimating.
- Group cows by yield, which results in faster milking.
- Evaluate long-term milk production for individual cows to plan breeding and culling.

Reproduction

Reproduction is a necessary and important part of milk production. Without regular calvings in the long run, there will be difficulties in producing the desired amount of milk. It is also important to produce sufficient heifer calves as replacement animals and allow herd size to be maintained or expanded.

Today it is most common to use artificial insemination (AI) instead of a bull. AI increases the control over the breeding and enables the dairy farmer to use sperm from all over the world, which increases the genetic gain. However, at the same time, it increases the need for good planning and well structured working routines.

Another way to get a cow pregnant is to use the embryo transfer technique. So far this is relatively uncommon, but it is expected to increase in importance. The main advantage with embryo transfer is that it is possible to generate more calves from a good cow than it is with AI.

The Oestrus Cycle

As long as a cow or heifer is not pregnant she will normally have a 21-day oestrus cycle. The length of the oestrus cycle may vary but it usually ranges from about 17 to 24 days. A heifers oestrus cycle is normally slightly shorter than a cows. The cycle will continue until the cow is pregnant. After calving, cows normally undergo a 20 to 30-day period when oestrus cycles do not occur.

The oestrus cycle is controlled by a complex system involving different hormones produced in the brain and ovary. The picture below shows a simplified picture of how two of these hormones, oestrogen and progesterone vary depending on where in the cycle the cow is.

Some cows do not follow the normal oestrus cycle. For example, a cow can be unoestrus, which means her ovaries do not function with the regular 17 to 24-day cycle and are therefore not observed in heat. Other cows may suffer from ovarian cysts. These cows will show heat at very short intervals and the period that they are in heat will last three to four days.

Reproductive Management

Research has shown that suboptimal calving interval causes large economic losses to dairy farmers, second in importance only to mastitis. Milking cows in late lactation are less profitable due to the decline in production. A long calving interval means milking less profitable cows due to the decline in production, fewer calves and too many cows with low feed conversion efficiencies. Therefore, successful reproductive management has a significant impact on the herds overall performance and the net income.

Well managed reproduction also reduces the risk of expensive involuntary culling. Information from UK and USA shows that the cost for a one day extended calving interval ranges from £1.5 to £3 per cow.

The time of the calving interval that is possible to influence by management is the open days, which are determined by the voluntary waiting period (VWP) and the breeding window (BW). A very common reason for undesired long calving intervals are missed heats. With improved heat detection rate (HDR) and conception rate (CR), through better management and improved timing of in seminations, it is possible to obtain a significantly shorter calving interval.

Detecting Heat

The most sexually intensive period of the oestrus cycle is during standing heat, which lasts for approximately 18 hours. In loose-housing herds, this period is indicated by the cow in heat standing immobile when mounted by another cow or bull. Other signs of heat are:

- Bellowing
- Increased activity
- Walking the fence line
- Licking/Sniffing
- Swelling and reddening of the vulva
- Mounting other cows
- Lower milk yield
- Reduced feed intake.

The duration of heat varies from animal to animal, but approximately 10 to 12 hours after the end of standing heat, the egg is released (ovulation) and the heat ends.

Manual Heat Detection

Manual heat detection relies on manual observations in the barn. The cows and heifers shall be observed for oestrus two to three times per day and all observed heats be recorded whether the animal is bred or not. The records provide the manager with information to anticipate future heats, which will make it easier to distinguish if a cow is in heat or not. Most mounting occurs between 6 p.m. and 6 a.m., and it is therefore worth trying to check for heat during these hours. To facilitate the planning and record keeping, a cow calendar is often used. This can be either manual or computerised.

The disadvantage with manual detection is that it is very time demanding and requires people with the ability to observe the right signs. This is particularly important when there are no distinct signs of heat. By using progesterone testing or other tools on the market, the detection rate can be improved.

Automatic Heat Detection

Another way to identify cows in heat is to monitor their activity. During heat the activity can increase up to 8 times compared to the normal level.

The activity can automatically be recorded by using activity meters attached to the neck or leg.

By comparing the activity with the last observed heat, actual milk yield and feed consumption, a reliable indicator of heat is obtained. The automation will generate a significant time saving and improved calving interval through better heat detection.

Timing of Insemination

With artificial insemination, the timing of the insemination becomes important. The optimum time for insemination depends on when the ovulation occurs in relation to the heat and for how long the sperm is viable. Most sperm remain viable for about 24 hours. The ovums life is only about four hours and is the most critical. It is therefore preferable that viable sperm is present in the salpin during ovulation.

There are mainly two rules for the timing of insemination. Traditionally the a.m.-p.m. rule was the one dairy producers followed. This rule dictates that cows and heifers first observed in heat in the morning should be bred late in the afternoon. Likewise, cows and heifers first observed in heat in the afternoon should be bred the following morning.

This is still a good rule, but many producers have now successfully gone to once-a-day insemination. This rule dictates that cows and heifers first observed in oestrus in the afternoon or the following morning, should be bred late that morning.

Herd Health

Disturbances and diseases will always occur in a herd, but in order to limit the economic loss, it is important to keep them under control. Good health and health management have a significant effect on a dairy farms net result. As the number of different disturbances is great, only the most common ones will be covered here.

Common Diseases and Disturbances

Retained placenta: This is a condition where the foetal membranes are not completely expelled. Retained placenta increases the risk for a whole host of other health problems (e.g., uterus infection, ketosis, and displaced abomasum). Usually a retained placenta is associated with difficult calving, a mineral imbalance (e.g., potassium, calcium) prior to calving, or over conditioning.

Milk Fever: Milk fever occurs when there is a shortage of calcium in the blood. This also usually results from a mineral imbalance (e.g., potassium, calcium) prior to calving. The effect of milk fever normally starts with reduced feed intake, followed by difficulties for the cow to move. Eventually the cow becomes paralysed and, if not treated, may die within a couple of hours. However, treatment with calcium usually has a good effect and in most cases, the cow will survive.

Ketosis: Ketosis occurs when cows begin to milk but have a shortage of energy. These cows will use fat to support milk production, but their liver cannot convert the fat to energy fast enough. Ketone bodies accumulate in the blood causing ketosis. Treatments generally consist of rapidly digested sources of glucose such as propylene glycol or molasses.

Left-Side Displacement of Abomasum (LDA): This may occur in relation to calving. The fourth stomach (abomasum) migrates from the right side of the cow to the left side. It is thought that this occurs with low fibre levels in the diet, physical stress such as slippery floors, and secondarily to other problems such as ketosis.

Acidosis: There are a few health problems directly associated with feeding, but high levels of concentrates might cause acidosis. The rumen pH drops below 6.0 for a long period of time causing numerous problems. One result of acidosis is hoof problems. Cows that have acidosis may experience hoof ulcers and a form of founder.

Mastitis: Mastitis is the most common and costly disease in dairy herds. It is an inflammation in the mammary gland which can be caused by bacterial infections or trauma. Cows with clinical mastitis are relatively easy to detect for the farmer. The symptoms are clotting and discoloration of the milk, and the gland becomes hard, red or swollen and in severe cases the cow has fever and loss of appetite. Cows with subclinical mastitis will have no outward signs of mastitis but their mammary gland will still be infected and the somatic cell count will be high.

Lameness: Hoof problems or lameness is characterised by damage to sensitive laminae of the hooves and it is reported to be the third most common cause of culling behind reproduction and mastitis. Hard and slippery floors, decreased exercise and metabolic disturbances (especially acidosis) negatively affect the occurrence of hoof problems. Estimates are that cattle that become lame and are not tended to, can experience a 20% loss in milk production.

Managing Herd Health

The economic losses due to health disturbances can be attributed to one or more of the following factors:

- Less efficient production and higher veterinary costs
- Reduced slaughter value and idle production factors
- Lost future income.

Lost future income occurs when the cows have to be replaced (culled) before reaching their economically optimal age. This varies from individual to individual, but it is basically the time when it is more profitable to replace a cow than to keep her. Culling is usually the eventual outcome of health problems and culling rates are normally found to be between 25.35%. Cows that die on the farm are part of this culling rate and should represent less than 3% of the total herd.

The reduction of losses attributed to diseases and the promotion of positive welfare and health should be one of the main goals in dairy herd management. A helpful tool for maintaining healthy cows is body condition scoring (BCS). Other important aspects that might affect health, if not looked after and maintained, are barn environment, milking routines and milking equipment.

Body Condition Scoring

Body condition scoring (BCS) can be used to troubleshoot problems and improve the health and productivity of the dairy herd. Overweight cows (overconditioned) are more susceptible to metabolic problems, infections and noninfectious health problems. Research shows that overweight cows are more likely to be affected by, for example, mastitis, retained placenta, ketosis and lameness.

They are also more likely to have difficulties at calving. Thinness (underconditioning) can lower the milk production and fat content because of insufficient energy and protein reserves.

The body condition scoring system uses a one to five, scale with one representing very thin cows and five very overweight cows. An ideal cow has a body condition score of about 3.5, but the system is designed to have cows at certain stages of lactation at certain body conditions. Important parts to look at when scoring are: chine, loin, rump, hip, tail head and pin bones.

Score	*Condition*
1	Skin and bones
2-2.5	Severe negative energy balance on cow in early lactation. Risk of production loss.
2.5-3	High producer in early lactation.
3-3.5	Milking cow in good nutrient balance.
3.5-4	Late lactation and dry cow in good condition.
4	Overconditioned. Potential calving problems if dry.
5	Severely overconditioned. Risk of fat cow syndrome.

Dry cows should neither gain nor lose weight and their body condition score should be around 3.5. If they have too much fat (>4.0) they tend to have many problems after calving. The reason is that all of this body fat must move through the liver before it can be used for milk production.

This is not bad if they use the fat slowly, but if they begin to lose weight quickly, fat will accumulate in the liver and a condition known as fatty liver develops. These cows usually perform poorly and are more prone to die. After calving, cows should lose less than one point before they begin to gain weight again. Cows that lose more than one point tend to have more reproductive problems.

From peak milk production to dry off, cows should gain back the body condition that was lost before the peak was reached. It is common for cows that have problems (e.g., ketosis, hoof problems, mastitis etc.) to lose one body condition score within the first two weeks of lactation. Ideally, the body condition score of a cow should be taken whenever she is handled. At least body condition scores should be taken at freshening, breeding and dry off. If over-or underconditioning occurs, the feeding strategy for an individual cow or group has to be evaluated and feed rations may have to be recalculated and corrected to satisfy the cow's demand.

Preventing and Controlling Mastitis

Mastitis needs extra attention, as it is the most common disease among dairy herds. It is not unusual that up to 40% of the herd is infected by clinical (with visible symptoms) or subclinical (often without symptoms) mastitis. All these cases are associated with many costs.

With good management routines, there are great opportunities to reduce the number of cases. This has a direct effect on production levels and costs. For example, if a farm is able to reduce the somatic cell count from 200 000 to 100 000, it means an increased average production of 0.7 kg per cow and day. In a 100 cow herd, this gives an additional milk production of 0.7 X 365 X 100 = 25 550 kg of milk per year.

The management must focus on actions that prevent new infections and reduces the duration of each infection. This includes proper maintenance of bedding and stalls, hygiene procedures during milking and frequent checking of the milking equipment. Dairy cows require a comfortable, clean, dry and draft-free environment. Comfort can include everything from feed space and walking surfaces to air and resting areas. However, even if the physical facilities are well planned, the daily work routines are of great importance to prevent as many health problems as possible. This includes not only mastitis, but also hoof problems (lameness). The way to avoid many of these cases is to regularly trim the hoofs and make sure that the surfaces in the barn are dry and free from mud.

Important Routines for Preventing Mastitis

With good milking routines and adequate milking equipment, the number of new mastitis cases will be significantly lowered. But even so, there will still be cases that need treatment. To avoid excessive costs associated with these cases, the duration of each case has to be reduced through:

- Prompt treatment of clinical cases; early detection, appropriate therapy and recording of clinical cases.
- Culling cows with chronic mastitis.
- Regularly checking the somatic cell count on all cows (High SCC indicates subclinical mastitis).
- Culling persistent offenders, e.g. cows having three cases in one quarter or five cases in mixed quarters in one lactation.

Milk Quality

Milk consumers require healthy and safe milk that is free from contamination, unpleasant smell etc. Processing also demands for milk to meet certain standards. For example, cheese making is very sensitive to spores and antibiotics. The requirements on the quality has increased over the time and will continue to do so. To meet this, more and more dairies put pressure on the milk producers. This is normally reflected in the payment scheme, where producers that deliver milk that does not meet the required standards have to pay penalties. There are also examples of processors paying bonuses to producers that meet the highest standards. As this makes the quality of the milk a determinant of the producers final price, it has to be considered in the daily routines.

The dairies use different parameters to check the quality of the milk. The exact level for each parameter varies from company to company. Therefore it is worth taking a closer look at the demands and standards you have to meet. Examples of tests carried out on delivered milk are:

- Fat and protein content
- Bacteria count (Bactoscan)
- Smell and taste
- Residues from antibiotics and other Medicines
- Freezing point (detects water content)
- Somatic Cell Count
- Spores (especially important for cheese making).

Managing Milk Quality

There are many factors that affect milk quality and some are easier to control than others. With well structured daily routines and an awareness among the personnel, it is possible to control and improve the situation considerably. Below you find a list of actions that all will have a positive effect on the overall quality and the final price of the milk:

- High quality water in sufficient amounts
- Feeds with high hygienic quality
- A well balanced ration with suitable ingredients
- Good overall hygiene
- A good and well maintained barn and parlour environment
- Prompt treatment of diseases
- Do not deliver colostrum milk
- Avoid delivering milk from cows with high cell count
- Do not deliver milk containing residues of antibiotics, sulpha etc.
- Do not add any water to the milk
- Make sure no detergent and/or other chemical residues get into the milk
- Ensure good ventilation to avoid undesired smell
- Make sure the milking equipment is cleaned properly
- Cool the milk directly after milking (Avoid freezing!).

4

Chemical Constituents of Milk

Composition of Milk

Milk from different animals varies somewhat in the proportion of the different constituents. All milk contains a high proportion of water, cow's milk averaging about 87 per cent. The milk from different breeds of cattle varies considerably in composition. The milk of individual cows of the same breed also varies in the proportion of the different constituents. This may be partly due to environment, partly to inheritance, and partly to individuality. Milk may vary particularly in fat content from one milking to the next. The percentage of fat increases during the milking period; that is, the first milk obtained is not so rich in fat as the last portions of milk.

The fat varies for different breeds from about 3.5 per cent for Holstein to about 5 per cent for Guernsey and Jersey.

The protein varies from about 3.3 per cent for Holstein to 4.0 per cent for Guernsey and Jersey. It parallels the fat content somewhat, being highest in the breeds having a high fat content and lowest in those having a low fat content.

The lactose does not vary so much with the different breeds as the fat and protein. The lactose content is from about 4.65 to 5 per cent.

The ash varies from about 0.68 to 0.75 for the different breeds.

The constituents of milk that are most important in food preparation are enzymes, vitamins, pigments, salts, sugar, fat, and proteins.

Enzymes. The enzymes of cow's milk are reported as follows by Rogers; proteinases, lactase, diastase, lipase, salolase, catalase, peroxidase, and aldehydrase. Rogers states that the proteolytic enzyme, galactase, brings about slow decomposition of milk proteins into peptones, amino acids, and ammonia.

Vitamins. All the vitamins recognized at the present time are contained in milk. Some are present in comparatively large and others in smaller amounts.

Pigments. The appearance of milk is white. This is due to light rays reflected by the colloidally dispersed constituents of the milk, the calcium caseinate, and calcium phosphate.

Milk contains two classes of yellow or orange pigments. The water-soluble pigment, which imparts a yellow colour with a green fluorescence to the whey of milk, was formerly called lactochrome. A name recently suggested for this pigment is lactoflavin. It is regarded as one flavin of a specific group, collectively to be called lyochromes. It is possible that lacto-flavin is composed of more than one pigment. Rogers says "lactoflavin forms compounds with saccharides, proteins, and purines (uric acid). These compounds possibly either occur naturally in milk or readily form during isolations, thus accounting for the several lactoflavins isolated from milk." It is probable that the pigment lactoflavin is one of the five fractions of vitamin G (B2). Milk is relatively rich in this vitamin.

A fat-soluble pigment, carotene, found in the fat gives the milk a more or less yellow tinge, which is more pronounced as the fat particles become more concentrated and form cream The group of pigments called caroti-noids, which include carotene, xanthophyll, and related pigments, has been described in the chapter on fruits and vegetables. The chief pigment of butter fat is the carotene, but little xanthophyll being found. The depth of colour depends upon the amount of pigment present. The colour of carotene in solution varies from yellow to orange and to a deep red-orange as the concentration increases. The amount of carotene found in the butter fat depends upon the extent of carotene in the food of the cow. Green grasses, hay cured to retain its green colour, green corn, and carrots are rich in carotene. The carotene content of milk fat is less rich during the winter months, if the food of the cow is poor in carotene during this period. Only in cow's milk is the fat extensively pigmented. With the exception of the fat of human milk, which is often pigmented, the fat of the milk of other animals is either devoid of or contains little pigment.

Salts. Milk contains salts of potassium, sodium, magnesium, calcium, phosphates, chlorides, and citrates. Traces of sulfates and carbonates are found. Iron is present in small amount. Iodides are also found in small amounts, the amount being greater in some localities than in others. Iodides may be easily transmitted from the feed to the milk. Supplee and Bellis have found copper to average about 0.52 part per million in freshly drawn milk. Brickner has reported that milk contains 3.6 to 5.6 parts of zinc per million parts of milk. Manganese in normal milk averages 0.02 to 0.06 parts per million. The greater part of the sulfur is found in the milk proteins. Barger and Coyne state that part of the sulfur in milk is found in the amino acids methionine and cystine of the proteins, but that all of the sulfur is not accounted for.

The salts of milk are found in milk in solution, in the colloidal state, and in combination with the proteins. The exact chemical combinations of the different salts in the milk are not fully determined. For this reason different authorities report different salt combinations. Thus formulas imply definite combination, whereas there is a complex salt equilibrium in milk, which has not been satisfactorily worked out. The citrates, the combinations of which may be trisodium and tripotassium citrate, tricalcium and trimagnesium citrate, are probably entirely in solution. The possible chlorides of potassium, sodium, and calcium are also in solution. Some authorities believe that the phosphates are present chiefly as dicalcium phosphate, CaHPO4 others believe that tricalcium phosphate, Ca3(PO4)2, is the principal phosphate combination. The phosphates are partly in solution, with the greater portion in colloidal dispersion. When the particles of dicalcium phosphate are heated they become aggregated and partially precipitated. Calcium and magnesium are in combination with the casein to form calcium and magnesium caseinates. Zoller states that there may be traces of sodium and potassium caseinates.

Lactose. The solubility of lactose and its properties may be found in the chapter on sugar. It is caramelized by heat at rather low temperatures.

Fat. Butter fat is composed of glycerol and fatty acids. Fatty acids of both the saturated and unsaturated series are present. The relative percentage of the unsaturated fatty acids varies with the feed, averaging higher in summer than in winter. Dean and Hilditch state that the oleic and linoleic acids increase by 4 per cent (mols), with a parallel diminution in butyric and stearic acids when cows return to pasture.

They also report that with increased age of the cow the unsaturated acids increase at the expense of palmitic acid, which was lowered from 29 to 22-23 per cent by weight of the total fat in the four years of observations made on milk from the same group of cows. The saturated fatty acids are butyric, caproic, caprylic, capric, lauric, myristic, plamitic, and stearic. Arachadonic acid has been reported absent by some investigators. Butter fat contains a higher proportion of the first-named saturated acids than other food fats. The first ones in the series are quite volatile with steam, the volatility decreasing with increase in molecular weight of the acid. Hence, when butter is heated for several minutes, the percentage of the lower saturated fatty acids may be decreased. The unsaturated fatty acids include oleic, linoleic, and arachidic.

Formerly it was thought that the particular flavour of butter was due to the greater proportion of the lower saturated fatty acids, but now it is known that butters with satisfactory flavour and aroma (Michaelian, Farmer, and Hammer) contain considerable quantities of acetylmethyl-carbinol plus diacetyl. The acetylmethylcarbinol in a pure condition is odorless, but by bacterial action it is changed to diacetyl, which in concentrations of 0.0002 to 0.0004 per cent in or added to neutral butter gives a characteristic aroma.

Size of Fat Globules

The fat in milk is found in small globules. These fat globules which are microscopic in size are suspended in the milk. They vary in size from 0.10μ to 22.0μ. Most of the globules are less than 10μ and average about 3μ in diameter. The size of the fat globules varies (1) with different breeds, being larger in milk from Jersey and Guernsey than in milk from other breeds, (2) with the lactation period, decreasing in size with length of the lactation period, and (3) with feed. Dry feed tends to decrease the size of the globules, succulent feed to increase their size.

Creaming

The specific gravity of the fat globules is less than that of the fluid of the milk. Hence there is a tendency for them to rise to the surface to form a cream layer. The extent of creaming depends upon several factors, such as the size of the fat globules, temperature of the milk, acidity, physical state of the fat, etc. Creaming occurs more rapidly in milk when the fat globules are quite large than when they are smaller. In rising to the top of the milk the globules of fat clump together, and this clumping increases the tendency for them to rise to the surface of the milk. As the larger clumps rise they may carry many of the smaller globules to the surface.

Therefore, clumping not only aids the completeness of creaming but also the rate, for the rate is more rapid when the fat particles clump quickly. Rogers states that the factors affecting clumping are "temperature, the acidity, the fat content and its degree of dispersion, the degree of agitation, and the fluidity of the system. The fat content, the degree of agitation and the fluidity of the system determine the probability of collisions of the globules." The tendency of the fat globules to clump is greatest when the milk is cooled rapidly to 7° to 8°C, but if the fat globules become solid at the low temperature before creaming is allowed to take place the rate of creaming is retarded. The temperature that favours clumping is also best for whipping the cream. Most cream is separated from the milk by mechanical means. The fat particles left in the milk after separating the cream are those less than 1u and those between 1 and 2u in diameter.

Butter

The fat globules in cow's milk are suspended in the milk and thus do not form a permanent emulsion, though they may be so reduced in size by homogenization that they form a permanent emulsion. Of the different theories formulated for explaining the manner in which emulsions are stabilized the adsorption film theory is usually connected with milk. Substances that lower surface tension tend to collect at the interface between two non-miscible systems. Proteins tend to lower the surface tension, hence tend to collect at the interface.

The layer or film around the fat globules probably consists of adsorbed calcium caseinate, with some lactal-bumin, globulin, and calcium phosphate. This membrane may be weakened or broken in various ways. When milk is heated slowly, the membrane surrounding some of the fat globules may be broken and a number of globules may coalesce. Sometimes milk that has been heated and then cooled has a more oily appearance due to this coalescing of the fat globules. The membrane surrounding the fat particles may be broken by mechanical agitation. Formation of butter in churning is brought about in this manner.

There are Two Theories Regarding Butter Formation: One is that the emulsion is reversed from the type found in the milk and that the butter is a water-in-fat emulsion. The other view is that butter is formed by packing the fat globules into a compact mass, and that water and air are enmeshed during this process.

Temperature and the formation of a foam are both important in churning. At a temperature above 65°C. there is no aggregation of fat particles. Below 65° the tendency to clump increases and is at a maximum at 7° to 8°C. A favourable temperature for butter formation is below 24° and above 10°C. At temperatures below 4°C. the fat globules do not adhere to each other and aggregation does not take place. At temperatures above the melting point of butter, no butter is formed.

The fat particles tend to clump at the liquid/air interface, so that air beaten in during the churning process accelerates clumping of the fat.

Butter may be churned from sweet or sour cream. The flavour of butter from the sweet cream is milder and different from that of the sour-cream butter. Many housekeepers prefer the sweet-cream butter for table use and the sour-cream butter for cooking.

The adsorbed films surrounding the fat globules may also be destroyed by the addition of acid or alkali. It is by these methods that the fat is set free for a quantitative determination. In the Babcock test, acid is used for liberating the fat globules; in the Hoyberg test alkali is used. The Bab-cock, or some modification of it, is the one usually employed in estimating the fat content of milk and cream.

Protein

The chief proteins found in milk in order of their decreasing amounts are casein, lactalbumin, and lactoglobulin.

Casein

Casein belongs to the group of phosphoproteins. The form in which the phosphorus exists in the casein is not definitely known, but it is believed to be present in the form of combined phosphoric acid. Casein forms about 3 per cent of cow's milk.

At its isoelectric point, which is pH 4.6, casein is nearly insoluble in water. Casein is amphoteric and forms salts with acids and alkalies. Fresh milk has a reaction of about pH 6.6, so that the casein is present in the milk as salts of bases and is found as calcium and magnesium caseinates. All the alkali caseinates are soluble in water, though the salts of the alkaline earths are less soluble than the alkali ones. Loeb states that below pH 4.6 the casein chloride, casein acetate, and casein lactate are very soluble in water, but casein sulfate and casein oxalate are difficultly soluble.

According to Zoller, pure casein when heated in water begins to imbibe water at 80° to 90°C. and becomes plastic. In this form it can be molded and shaped. Upon cooling it becomes very hard. Casein can be precipitated from milk by bringing the milk to the isoelectric point of casein. Coagulation of casein will be considered later.

Lactalbumin

The proportion of lactalbumin in milk is much lower than that of casein. It forms about 0.50 per cent of cow's milk. Its isoelectric point is pH 4.55. Since the reaction of fresh milk is about pH 6.6, it is on the alkaline side of the isoelectric point of lactalbumin.

Thus is it possible that the lactalbumin is found combined as salts of bases, such as calcium and magnesium albuminates. Osborne and Wakeman think it is uncom-bined. Lactalbumin is soluble in water, and is coagulated by heating in solution to a temperature of about 70°C. Coagulation may not be complete at this temperature. Palmer states that the lactalbumin is more highly dispersed than the other colloidal constituents of the milk.

Lactoglobulin

Lactoglobulin occurs in milk in very small quantities, about 0.05 per cent of cow's milk. Lactoglobulin is insoluble in distilled water, but it is soluble in dilute solutions of strong bases or acids, and in dilute salt solutions. It is coagulable by heat.

Physical and Chemical Properties of Milk

The role of milk in nature is to nourish and provide immunological protection for the mammalian young. Milk has been a food source for humans since prehistoric times; from human, goat, buffalo, sheep, yak, to the focus of this section-domesticated cow milk (genus Bos). Milk and honey are the only articles of diet whose sole function in nature is food. It is not surprising, therefore, that the nutritional value of milk is high.

Milk is also a very complex food with over 100,000 different molecular species found. There are many factors that can affect milk composition such as breed variations (see introduction, cow to cow variations, herd to herd variations-including management and feed considerations, seasonal variations, and geographic variations. With all this in mind, only an approximate composition of milk can be given:

- 87.3% water (range of 85.5%-88.7%)
- 3.9 % milkfat (range of 2.4%-5.5%)
- 8.8% solids-not-fat (range of 7.9-10.0%):
  - o protein 3.25% (3/4 casein)
  - o lactose 4.6%
  - o minerals 0.65%-Ca, P, citrate, Mg, K, Na, Zn, Cl, Fe, Cu, sulfate, bicarbonate, many others
  - o acids 0.18%-citrate, formate, acetate, lactate, oxalate
  - o enzymes-peroxidase, catalase, phosphatase, lipase
  - o gases-oxygen, nitrogen
  - o vitamins-A, C, D, thiamine, riboflavin, others.

The following terms are used to describe milk fractions:

- Plasma = milk-fat (skim milk)
- Serum = plasma-casein micelles (whey)
- solids-not-fat (SNF) = proteins, lactose, minerals, acids, enzymes, vitamins
- Total Milk Solids = fat + SNF.

Not only is the composition important in determining the properties of milk, but the physical structure must also be examined. Due to its role in nature, milk is in a liquid form. This may seem curious if one takes into consideration the fact that milk has less water than most fruits and vegetables. Milk can be described as:

- an oil-in-water emulsion with the fat globules dispersed in the continuous serum phase
- a colloid suspension of casein micelles, globular proteins and lipoprotein partilcles
- a solution of lactose, soluble proteins, minerals, vitamins other components.

Looking at milk under a microscope, at low magnification (5X) a uniform but turbid liquid is observed. At 500X magnification, spherical droplets of fat, known as fat globules, can be seen. At even higher magnification (50,000X), the casein micelles can be observed. The main structural components of milk, fat globules and casein micelles, will be examined in more detail later.

Caseins

The casein content of milk represents about 80% of milk proteins. The principal casein fractions are alpha(s1) and alpha(s2)-caseins, ß-casein, and kappa-casein. The distinguishing property of all caseins is their low solubility at pH 4.6. The common compositional factor is that caseins are conjugated proteins, most with phosphate group(s) esterified to serine residues. These phosphate groups are important to the structure of the casein micelle. Calcium binding by the individual caseins is proportional to the phosphate content.

The conformation of caseins is much like that of denatured globular proteins. The high number of proline residues in caseins causes particular bending of the protein chain and inhibits the formation of close-packed, ordered secondary structures. Caseins contain no disulfide bonds. As well, the lack of tertiary structure accounts for the stability of caseins against heat denaturation because there is very little structure to unfold. Without a tertiary structure there is considerable exposure of hydrophobic residues. This results in strong association reactions of the caseins and renders them insoluble in water.

Within the group of caseins, there are several distinguishing features based on their charge distribution and sensitivity to calcium precipitation:

Alpha(s1)-casein: (molecular weight 23,000; 199 residues, 17 proline residues) Two hydrophobic regions, containing all the proline residues, separated by a polar region, which contains all but one of eight phosphate groups. It can be precipitated at very low levels of calcium.

Alpha(s2)-casein: (molecular weight 25,000; 207 residues, 10 prolines) Concentrated negative charges near N-terminus and positive charges near C-terminus. It can also be precipitated at very low levels of calcium.

B-casein: (molecular weight 24,000; 209 residues, 35 prolines) Highly charged N-terminal region and a hydrophobic C-terminal region. Very amphiphilic protein acts like a detergent molecule. Self association is temperature dependant; will form a large polymer at 20° C but not at 4° C. Less sensitive to calcium precipitation.

Kappa-casein: (molecular weight 19,000; 169 residues, 20 prolines) Very resistant to calcium precipitation, stabilizing other caseins. Rennet cleavage at the Phe105-Met106 bond eliminates the stabilizing ability, leaving a hydrophobic portion, para-kappa-casein, and a hydrophilic portion called kappa-casein glycomacropeptide (GMP), or more accurately, caseinomacropeptide (CMP).

Structure: The Casein Micelle

Most, but not all, of the casein proteins exist in a colloidal particle known as the casein micelle. Its biological function is to carry large amounts of highly insoluble CaP to mammalian young in liquid form and to form a clot in the stomach for more efficient nutrition. Besides casein protein, calcium and phosphate, the micelle also contains citrate, minor ions, lipase and plasmin enzymes, and entrapped milk serum.

These micelles are rather porous structures, occupying about 4 ml/g and 6-12% of the total volume fraction of milk. The "casein sub-micelle" model has been prominent for the last several years, and is illustrated and described with the following link, but there is not universal acceptance of this model, and mounting research evidence to suggest that there is not a defined sub-micellar structure to the micelle at all. Another model of a more open structure is also defined with the following link.

In the submicelle model, it is thought that there are small aggregates of whole casein, containing 10 to 100 casein molecules, called submicelles. It is thought that there are two different kinds of submicelle; with and without kappa-casein. These submicelles contain a hydrophobic core and are covered by a hydrophilic coat which is at least partly comprised of the polar moieties of kappa-casein. The hydrophilic CMP of the kappa-casein exists as a flexible hair.

The open model also suggests there are more dense and less dense regions within the midelle, but there is less of a well-defined structure. In this model, calcium phosphate nanoclusters bind caseins and provide for the differences in density within the casein micelle.

Colloidal calcium phosphate (CCP) acts as a cement between the hundreds or even thousands of submicelles that form the casein micelle. Binding may be covalent or electrostatic. Submicelles rich in kappa-casein occupy a surface position, whereas those with less are buried in the interior. The resulting hairy layer, at least 7 nm thick, acts to prohibit further aggregation of submicelles by steric repulsion. The casein micelles are not static; there are three dynamic equilibria between the micelle and its surroundings:

- the free casein molecules and submicelles
- the free submicelles and micelles
- the dissoved colloidal calcium and phosphate.

The following factors must be considered when assessing the stability of the casein micelle:

Role of Ca++: More than 90% of the calcium content of skim milk is associated in some way or another with the casein micelle. The removal of Ca++ leads to reversible dissociation of ß-casein without micellular disintegration. The addition of Ca++ leads to aggregation.

H Bonding: Some occurs between the individual caseins in the micelle but not much because there is no secondary structure in casein proteins.

Disulphide Bonds: alpha(s1) and ß-caseins do not have any cysteine residues. If any S-S bonds occur within the micelle, they are not the driving force for stabilization.

Hydrophobic Interactions: Caseins are among the most hydrophobic proteins and there is some evidence to suggest they play a role in the stability of the micelle. It must be remembered that hydrophobic interactions are very temperature sensitive.

Electrostatic Interactions: Some of the subunit interactions may be the result of ionic bonding, but the overall micellar structure is very loose and open.

Van der Waals Forces: No sucess in relating these forces to micellular stability.

Steric stabilization: As already noted, the hairy layer interferes with interparticle approach.

There are several factors that will affect the stability of the casein micelle system:

Salt content: affects the calcium activity in the serum and calcium phosphate content of the micelles.

pH: lowering the pH leads to dissolution of calcium phosphate until, at the isoelectric point (pH 4.6), all phosphate is dissolved and the caseins precipitate.

Temperature: at 4° C, beta-casein begins to dissociate from the micelle, at 0° C, there is no micellar aggregation; freezing produces a precipitate called cryo-casein.

Heat Treatment: whey proteins become adsorbed, altering the behaviour of the micelle.

Dehydration: by ethanol, for example, leads to aggregation of the micelles.

When two or more of these factors are applied together, the effect can also be additive.

Casein Micelle Aggregation

Caseins are able to aggregate if the surface of the micelle is reactive. The Schmidt model further illustrates this.

Although the casein micelle is fairly stable, there are four major ways in which aggregation can be induced:

1. chymosin-rennet or other proteolytic enzymes as in Cheese manufacturing
2. acid
3. heat
4. age gelation.

Enzyme Coagulation

Chymosin, or rennet, is most often used for enzyme coagulation. During the primary stage, rennet cleaves the Phe(105)-Met(106) linkage of kappa-casein resulting in the formation of the soluble CMP which diffuses away from the micelle and para-kappa-casein, a distinctly hydrophobic peptide that remains on the micelle. The patch or reactive site, as illustrated in the above image, that is left on the micelles after enzymatic cleavage is necessary before aggregation of the paracasein micelles can begin.

During the secondary stage, the micelles aggregate. This is due to the loss of steric repulsion of the kappa-casein as well as the loss of electrostatic repulsion due to the decrease in pH. As the pH approaches its isoelectric point (pH 4.6), the caseins aggregate.

The casein micelles also have a strong tendency to aggregate because of hydrophobic interactions. Calcium assists coagulation by creating isoelctric conditions and by acting as a bridge between micelles. The temperature at the time of coagulation is very important to both the primary and secondary stages. With an increase in temperature up to 40° C, the rate of the rennet reaction increases. During the secondary stage, increased temperatures increase the hydrophobic reaction. The tertiary stage of coagulation involves the rearrangement of micelles after a gel has formed. There is a loss of paracasein identity as the milk curd firms and syneresis begins.

Acid Coagulation

Acidification causes the casein micelles to destabilize or aggregate by decreasing their electric charge to that of the isoelectric point. At the same time, the acidity of the medium increases the solubility of minerals so that organic calcium and phosphorus contained in the micelle gradually become soluble in the aqueous phase. Casein micelles disintegrate and casein precipitates. Aggregation occurs as a result of entropically driven hydrophobic interactions.

Heat

At temperatures above the boiling point casein micelles will irreversibly aggregate. On heating, the buffer capacity of milk salts change, carbon dioxide is released, organic acids are produced, and tricalcium phophate and casein phosphate may be precipitated with the release of hydrogen ions.

Age Gelation

Age gelation is an aggregation phenomenon that affects shelf-stable, sterilized dairy products, such as concentrated milk and UHT milk products. After weeks to months storage of these products, there is a sudden sharp increase in viscosity accompanied by visible gelation and irreversible aggregation of the micelles into long chains forming a three-dimensional network. The actual cause and mechanism is not yet clear, however, some theories exist:

1. Proteolytic breakdown of the casein: bacterial or native plasmin enzymes that are resistant to heat treatment may lead to the formation of a gel
2. Chemical reactions: polymerization of casein and whey proteins due to Maillard type or other chemical reactions
3. Formation of kappa-casein-ß-lactoglobulin complexes

An excellent source of information on casein micelle stability can be found in Walstra.

Whey Proteins

The proteins appearing in the supernatant of milk after precipitation at pH 4.6 are collectively called whey proteins. These globular proteins are more water soluble than caseins and are subject to heat denaturation. Native whey proteins have good gelling and whipping properties. Denaturation increases their water holding capacity. The principle fractions are ß-lactoglobulin, alpha-lactalbumin, bovine serum albumin (BSA), and immunoglobulins (Ig).

ß-Lactoglobulins: (MW-18,000; 162 residues) This group, including eight genetic variants, comprises approximately half the total whey proteins. ß-Lactoglobulin has two internal disulfide bonds and one free thiol group. The conformation includes considerable secondary structure and exists naturally as a noncovalent linked dimer. At the isoelectric point (pH 3.5 to 5.2), the dimers are further associated to octamers but at pH below 3.4, they are dissociated to monomers.

Alpha-Lactalbumins: (MW-14,000; 123 residues) These proteins contain eight cysteine groups, all involved in internal disulfide bonds, and four tryptophan residues. alpha-Lactalbumin has a highly ordered secondary structure, and a compact, spherical tertiary structure. Thermal denaturation and pH <4.0 results in the release of bound calcium.

Enzymes

Enzymes are a group of proteins that have the ability to catalyze chemical reactions and the speed of such reactions. The action of enzymes is very specific. Milk contains both indigenous and exogenous enzymes. Exogenous enzymes mainly consist of heat-stable enzymes produced by psychrotrophic bacteria: lipases, and proteinases. There are many indigenous enzymes that have been isolated from milk. The most significant group are the hydrolases:

- lipoprotein lipase
- plasmin
- alkaline phosphatase.

Lipoprotein lipase (LPL): A lipase enzyme splits fats into glycerol and free fatty acids. This enzyme is found mainly in the plasma in association with casein micelles. The milkfat is protected from its action by the FGM. If the FGM has been damaged, or if certain cofactors (blood serum lipoproteins) are present, the LPL is able to attack the lipoproteins of the FGM. Lipolysis may be caused in this way.

Plasmin: Plasmin is a proteolytic enzyme; it splits proteins. Plasmin attacks both ß-casein and alpha(s2)-casein. It is very heat stable and responsible for the development of bitterness in pasteurized milk and UHT processed milk. It may also play a role in the ripening and flavour development of certain cheeses, such as Swiss cheese.

Alkaline phosphatase: Phosphatase enzymes are able to split specific phosporic acid esters into phosphoric acid and the related alcohols. Unlike most milk enzymes, it has a pH and temperature optima differing from physiological values; pH of 9.8. The enzyme is destroyed by minimum pasteurization temperatures, therefore, a phosphatase test can be done to ensure proper pasteurization.

Lactose

Lactose is a disaccharide (2 sugars) made up of glucose and galactose (which are both monosaccharides). Because of the anomeric carbon on the right side of the structure below, lactose can exist as two isomers, alpha, as shown, or beta, in which the hydroxyl on the anomeric carbon would point up on the ring structure shown below.

It comprises 4.8 to 5.2% of milk, 52% of milk SNF, and 70% of whey solids. It is not as sweet as sucrose. When lactose is hydrolyzed by ß-D-galactosidase (lactase), an enzyme that splits these monosaccharides, the result is increased sweetness, and depressed freezing point.

One of its most important functions is its utilization as a fermentation substrate. Lactic acid bacteria produce lactic acid from lactose, which is the beginning of many fermented dairy products. Because of their ability to metabolize lactose, they have a competitive advantage over many pathogenic and spoilage organisms.

Some people suffer from lactose intolerance; they lack the lactase enzyme, hence they cannot digest lactose, or dairy products containing lactose. Crystallization of lactose occurs in an alpha form which commonly takes a tomahawk shape.

This results in the defect called sandiness. Lactose is relatively insoluble which is a problem in many dairy products, ice cream, sweetened condensed milk. In addition to lactose, fresh milk contains other carbohydrates in small amounts, including glucose, galactose, and oligosaccharides.

Vitamins

Vitamins are organic substances essential for many life processes. Milk includes fat soluble vitamins A, D, E, and K. Vitamin A is derived from retinol and ß-carotene. Because milk is an important source of dietary vitamin A, fat reduced products which have lost vitamin A with the fat, are required to supplement the product with vitamin A.

Milk is also an important source of dietary water soluble vitamins:

- B1-thiamine
- B2-riboflavin
- B6-pyridoxine
- B12-cyanocobalamin
- niacin
- pantothenic acid.

There is also a small amount of vitamin C (ascorbic acid) present in raw milk but it is an insignificant amount relative to human needs and is quite heat-labile: about 20% is destroyed by pasteurization.

The vitamin content of fresh milk is given below:

Vitamin	***Contents per litre***
A (ug RE)	400
D (IU)	40
E (ug)	1000
K (ug)	50
B1 (ug)	450
B2 (ug)	1750
Niacin (ug)	900
B6 (ug)	500
Pantothenic acid (ug)	3500
Biotin (ug)	35
Folic acid (ug)	55
B12 (ug)	4.5
C (mg)	20

Minerals

All 22 minerals considered to be essential to the human diet are present in milk. These include three families of salts:

1. *Sodium (Na), Potassium (K) and Chloride (Cl):These* free *ions are negatively correlated to lactose to maintain osmotic equilibrium of milk with blood.*
2. *Calcium (Ca), Magnesium (Mg), Inorganic Phosphorous (P(i)), and Citrate: This group consists of 2/3 of the Ca, 1/3 of the Mg, 1/2 of the P(i), and less than 1/10 of the citrate in* colloidal *(nondiffusible) form and present in the casein micelle.*
3. Diffusible *salts of Ca, Mg, citrate, and phosphate: These salts are very pH dependent and contribute to the overall acid-base equilibrium of milk.*

The mineral content of fresh milk is given below. A wide range is shown, based on variability due to intrinsic (lactogenesis in the secretory cell of individual animals) and extrinsic factors (mainly feed and season/climate adaptation, but also contaminants).

Mineral	*Content per litre*
Sodium (mg)	250-640
Potassium (mg)	1100-1500
Chloride (mg)	800-1200
Calcium (mg)	1100-1300
Magnesium (mg)	70-140
Phosphorus (mg)	800-1000
Iron (ug)	100-700
Zinc (ug)	2500-7000
Copper (ug)	100-350
Manganese (ug)	10-50
Iodine (ug)	50-600
Fluoride (ug)	20-80
Selenium (ug)	20-40
Cobalt (ug)	0.5-1.3
Chromium (ug)	0.5-20
Molybdenum (ug)	20-100
Nickel (ug)	0-50
Arsenic (ug)	20-60
Aluminum (ug)	50-1600
Lead (ug)	tr-20

This information is consolidated from C. D. Hunt and F. H. Nielsen, P.L.H. and P.F. Fox, 2009. Advanced Dairy Chemistry, Vol. 3, Lactose, water, salts and minor constituents, 3rd edn.,

Springer, pp. 392-8; and from, Cristina Sola-Larranaga; Inigo Navarro-Blasco. 2009. Chemometric analysis of minerals and trace elements in raw cow milk from the community of Navarra, Spain. Food Chemistry, 112 (1), 189-196.

5

The More Common Dairy Processes

The dairy industry is divided into two main production areas:

- the primary production of milk on farms—the keeping of cows (and other animals such as goats, sheep etc.) for the production of milk for human consumption;
- the processing of milk—with the objective of extending its saleable life. This objective is typically achieved by (a) heat treatment to ensure that milk is safe for human consumption and has an extended keeping quality, and (b) preparing a variety of dairy products in a semi-dehydrated or dehydrated form (butter, hard cheese and milk powders), which can be stored.

The focus of this document is on the processing of milk and the production of milk-derived products—butter, cheese and milk powder— at dairy processing plants. The upstream process of primary milk production on dairy farms is not covered, since this activity is more related to the agricultural sector. Similarly, downstream processes of distribution and retail are not covered.

Dairy processing occurs world-wide; however the structure of the industry varies from country to country. In less developed countries, milk is generally sold directly to the public, but in major milk producing countries most milk is sold on a wholesale basis. In Ireland and Australia, for example, many of the large-scale processors are owned by the farmers as co-operatives, while in the United States individual contracts are agreed between farmers and processors.

Dairy processing industries in the major dairy producing countries have undergone rationalisation, with a trend towards fewer but larger plants operated by fewer people. As a result, in the United States, Europe, Australia and New Zealand most dairy processing plants are quite large. Plants producing market milk and products with short shelf life, such as yogurts, creams and soft cheeses, tend to be located on the fringe of urban centres close to consumer markets. Plants manufacturing items with longer shelf life, such as butter, milk powders, cheese and whey powders, tend to be located in rural areas closer to the milk supply.

The general tendency world-wide, is towards large processing plants specialising in a limited range of products. There are exceptions, however. In eastern Europe for example, due to the former supply-driven concept of the market, it is still very common for 'city' processing plants to be large multi-product plants producing a wide range of products. The general trend towards large processing plants has provided companies with the opportunity to acquire bigger, more automated and more efficient equipment. This technological development has, however, tended to increase environmental loadings in some areas due to the requirement for long-distance distribution.

Basic dairy processes have changed little in the past decade. Specialised processes such as ultrafiltration (UF), and modern drying processes, have increased the opportunity for the recovery of milk solids that were formerly discharged. In addition, all processes have become much more energy efficient and the use of electronic control systems has allowed improved processing effectiveness and cost savings.

Process Overview

Milk Production

The processes taking place at a typical milk plant include:

- receipt and filtration/clarification of the raw milk;
- separation of all or part of the milk fat (for standardisation of market milk, production of cream and butter and other fat-based products, and production of milk powders);
- pasteurisation;
- homogenisation (if required);
- deodorisation (if required);
- further product-specific processing;
- packaging and storage, including cold storage for perishable products;
- distribution of final products.

In such plants, yogurts and other cultured products may also be produced from whole milk and skimmed milk.

Butter Production

The butter-making process, whether by batch or continuous methods, consists of the following steps:

- preparation of the cream;
- destabilisation and breakdown of the fat and water emulsion;
- aggregation and concentration of the fat particles;
- formation of a stable emulsion;
- packaging and storage;
- distribution.

Is a flow diagram outlining the basic processing system for a butter-making plant. The initial steps, (filtration/clarification, separation and pasteurisation of the milk) are the same as described in the previous section. Milk destined for butter making must not be homogenised, because the cream must remain in a separate phase. After separation, cream to be used for butter making is heat treated and cooled under conditions that facilitate good whipping and churning. It may then be ripened with a culture that increases the content of diacetyl, the compound responsible for the flavour of butter.

Alternatively, culture inoculation may take place during churning. Butter which is flavour enhanced using this process is termed lactic, ripened or cultured butter. This process is very common in continental European countries. Although the product is claimed to have a superior flavour, the storage life is limited. Butter made without the addition of a culture is called sweet cream butter. Most butter made in the English-speaking world is of this nature.

Both cultured and sweet cream butter can be produced with or without the addition of salt. The presence of salt affects both the flavour and the keeping quality. Butter is usually packaged in bulk quantities (25 kg) for long-term storage and then re-packed into marketable portions (usually 250 g or 500 g, and single-serve packs of 10–15 g). Butter may also be packed in internally lacquered cans, for special markets such as the tropics and the Middle East.

Cheese Production

Virtually all cheese is made by coagulating milk protein (casein) in a manner that traps milk solids and milk fat into a curd matrix. This curd matrix is then consolidated to express the liquid fraction, cheese whey. Cheese whey contains those milk solids which are not held in the curd mass, in particular most of the milk sugar (lactose) and a number of soluble proteins.

Milk Powder Production

Milk used for making milk powder, whether it be whole or skim milk, is not pasteurised before use. The milk is preheated in tubular heat exchangers before being dried. The preheating temperature depends on the season (which affects the stability of the protein in the milk) and on the characteristics desired for the final powder product.

The preheated milk is fed to an evaporator to increase the concentration of total solids. The solids concentration that can be reached depends on the efficiency of the equipment and the amount of heat that can be applied without unduly degrading the milk protein.

The milk concentrate is then pumped to the atomiser of a drying chamber. In the drying chamber the milk is dispersed as a fine fog-like mist into a rapidly moving hot air stream, which causes the individual mist droplets to instantly evaporate. Milk powder falls to the bottom of the chamber, from where it is removed. Finer milk powder particles are carried out of the chamber along with the hot air stream and collected in cyclone separators.

Milk powders are normally packed and distributed in bulk containers or in 25 kg paper packaging systems. Products sold to the consumer market are normally packaged in cans under nitrogen. This packaging system improves the keeping quality, especially for products with high fat content.

Environmental Impacts

This section briefly describes some of the environmental impacts associated with the primary production of milk and the subsequent processing of dairy products. While it is recognised that the primary production of milk has some significant environmental impacts, this document is predominantly concerned with the processing of dairy products.

Impacts of Primary Production

The main environmental issues associated with dairy farming are:

- the generation of solid manure and manure slurries, which may pollute surface water and groundwater;
- the use of chemical fertilisers and pesticides in the production of pastures and fodder crops, which may pollute surface water and groundwater;
- the contamination of milk with pesticides, antibiotics and other chemical residues.

In most cases, solid manure is applied to pastures and cultivated land. The extent of application, however, may be restricted in some regions. Dairy effluent and slurries are generally held in some form of lagoon to allow sedimentation and biological degradation before they are irrigated onto land. Sludge generated from biological treatment of the dairy effluent can also be applied to pastures, as long as it is within the allowable concentrations for specified pollutants, as prescribed by regulations. Sludge can also be used in the production of methane-rich biogas, which can then be used to supplement energy supplies. Manure waste represents a valuable source of nutrients. However improper storage and land application of manure and slurries can result in serious pollution of surface waters and groundwater, potentially contaminating drinking water supplies. The extensive use of chemical fertilisers containing high levels of nitrogen has resulted in pollution of the groundwater and surface waters in many countries.

Nitrite in drinking water is known to be carcinogenic, and nitrite levels in drinking water that exceed 25–50 mg/L have been linked to cyanosis in newborn infants ('blue babies').

Compounds containing nitrogen and phosphorus, if discharged to surface water, can lead to excessive algal growth (eutrophication). This results in depleted dissolved oxygen levels in the water, thereby causing the death of fish and other aquatic species. In sensitive areas, therefore, the rate and manner of application of chemical fertilisers are critical.

The use of pesticides has been recognised as an environmental concern for many agricultural activities. Toxic pesticides, some of which biodegrade very slowly, can accumulate in body tissues and are harmful to ecosystems and to human health. Pesticides can end up in agricultural products, groundwater and surface waters, and in extreme cases can enter the human food chain through milk.

For the past few decades, the contamination of milk with antibiotics has been an issue of concern. This is due to the overuse of antibiotics for treatment of cattle diseases, particularly mastitis. It has been brought under control in most countries with developed dairy industries, through strict limitations on the use of antibiotics, regular testing of milk for antibiotic residues, rigorous enforcement of regulations, and education.

In some countries, considerable attention has also been paid to the screening of milk supplies for traces of radioactivity, and most countries now apply acceptance limits for raw and imported milk products. Even the slightest levels of contamination in milk can be serious, because pollutants are concentrated in the processing process.

Impacts of Dairy Processing

As for many other food processing operations, the main environmental impacts associated with all dairy processing activities are the high consumption of water, the discharge of effluent with high organic loads and the consumption of energy. Noise, odour and solid wastes may also be concerns for some plants.

Dairy processing characteristically requires very large quantities of fresh water. Water is used primarily for cleaning process equipment and work areas to maintain hygiene standards.

The dominant environmental problem caused by dairy processing is the discharge of large quantities of liquid effluent. Dairy processing effluents generally exhibit the following properties:

- high organic load due to the presence of milk components;
- fluctuations in pH due to the presence of caustic and acidic cleaning agents and other chemicals;
- high levels of nitrogen and phosphorus;
- fluctuations in temperature.

If whey from the cheese-making process is not used as a by-product and discharged along with other wastewaters, the organic load of the resulting effluent is further increased, exacerbating the environmental problems that can result.

In order to understand the environmental impact of dairy processing effluent, it is useful to briefly consider the nature of milk. Milk is a complex biological fluid that consists of water, milk fat, a number of proteins (both in suspension and in solution), milk sugar (lactose) and mineral salts.

Dairy products contain all or some of the milk constituents and, depending on the nature and type of product and the method of manufacturing, may also contain sugar, salts (e.g. sodium chloride), flavours, emulsifiers and stabilisers.

For plants located near urban areas, effluent is often discharged to municipal sewage treatment systems. For some municipalities, the effluent from local dairy processing plants can represent a significant load on sewage treatment plants. In extreme cases, the organic load of waste milk solids entering a sewage system may well exceed that of the township's domestic waste, overloading the system.

In rural areas, dairy processing effluent may also be irrigated to land. If not managed correctly, dissolved salts contained in the effluent can adversely affect soil structure and cause salinity. Contaminants in the effluent can also leach into underlying groundwater and affect its quality.

In some locations, effluent may be discharged directly into water bodies. However this is generally discouraged as it can have a very negative impact on water quality due to the high levels of organic matter and resultant depletion of oxygen levels.

Electricity is used for the operation of machinery, refrigeration, ventilation, lighting and the production of compressed air. Like water consumption, the use of energy for cooling and refrigeration is important for ensuring good keeping quality of dairy products and storage temperatures are often specified by regulation. Thermal energy, in the form of steam, is used for heating and cleaning. As well as depleting fossil fuel resources, the consumption of energy causes air pollution and greenhouse gas emissions, which have been linked to global warming.

Dairy products such as milk, cream and yogurt are typically packed in plastic-lined paperboard cartons, plastic bottles and cups, plastic bags or reusable glass bottles. Other products, such as butter and cheese, are wrapped in foil, plastic film or small plastic containers. Milk powders are commonly packaged in multi-layer kraft paper sacs or tinned steel cans, and some other products, such as condensed milks, are commonly packed in cans.

Breakages and packaging mistakes cannot be totally avoided. Improperly packaged dairy product can often be returned for reprocessing; however the packaging material is generally discarded.

Emissions to air from dairy processing plants are caused by the high levels of energy consumption necessary for production. Steam, which is used for heat treatment processes (pasteurisation, sterilisation, drying etc.) is generally produced in on-site boilers, and electricity used for cooling and equipment operation is purchased from the grid. Air pollutants, including oxides of nitrogen and sulphur and suspended particulate matter, are formed from the combustion of fossil fuels, which are used to produce both these energy sources. In addition, discharges of milk powder from the exhausts of spray drying equipment can be deposited on surrounding surfaces. When wet these deposits become acidic and can, in extreme cases, cause corrosion.

For operations that use refrigeration systems based on chlorofluorocarbons (CFCs), the fugitive loss of these gases to the atmosphere is an important environmental consideration, since CFCs are recognised to be a cause of ozone depletion in the atmosphere. For such operations, the replacement of CFC-based systems with non-or reduced-CFC systems is thus an important issue.

Some processes, such as the production of dried casein, require the use of hammer mills to grind the product. The constant noise generated by this equipment has been known to be a nuisance in surrounding residential areas. The use of steam injection for heat treatment of milk and for the creation of reduced pressure in evaporation processes also causes high noise levels.

A substantial traffic load in the immediate vicinity of a dairy plant is generally unavoidable due to the regular delivery of milk (which may be on a 24-hour basis), deliveries of packaging and the regular shipment of products.

Noise problems should be taken into consideration when determining plant location. Hazardous wastes consist of oily sludge from gearboxes of moving machines, laboratory waste, cooling agents, oily paper filters, batteries, paint cans etc. At present, in western Europe some of these materials are collected by waste companies. While some waste is incinerated, much is simply dumped. Temperatures are often specified by regulation.

Thermal energy, in the form of steam, is used for heating and cleaning. As well as depleting fossil fuel resources, the consumption of energy causes air pollution and greenhouse gas emissions, which have been linked to global warming.

Dairy products such as milk, cream and yogurt are typically packed in plastic-lined paperboard cartons, plastic bottles and cups, plastic bags or reusable glass bottles. Other products, such as butter and cheese, are wrapped in foil, plastic film or small plastic containers. Milk powders are commonly packaged in multi-layer kraft paper sacs or tinned steel cans, and some other products, such as condensed milks, are commonly packed in cans.

Breakages and packaging mistakes cannot be totally avoided. Improperly packaged dairy product can often be returned for reprocessing; however the packaging material is generally discarded.

Emissions to air from dairy processing plants are caused by the high levels of energy consumption necessary for production. Steam, which is used for heat treatment processes (pasteurisation, sterilisation, drying etc.) is generally produced in on-site boilers, and electricity used for cooling and equipment operation is purchased from the grid. Air pollutants, including oxides of nitrogen and sulphur and suspended particulate matter, are formed from the combustion of fossil fuels, which are used to produce both these energy sources.

In addition, discharges of milk powder from the exhausts of spray drying equipment can be deposited on surrounding surfaces. When wet these deposits become acidic and can, in extreme cases, cause corrosion.

For operations that use refrigeration systems based on chlorofluorocarbons (CFCs), the fugitive loss of these gases to the atmosphere is an important environmental consideration, since CFCs are recognised to be a cause of ozone depletion in the atmosphere. For such operations, the replacement of CFC-based systems with non-or reduced-CFC systems is thus an important issue.

Some processes, such as the production of dried casein, require the use of hammer mills to grind the product. The constant noise generated by this equipment has been known to be a nuisance in surrounding residential areas. The use of steam injection for heat treatment of milk and for the creation of reduced pressure in evaporation processes also causes high noise levels.

A substantial traffic load in the immediate vicinity of a dairy plant is generally unavoidable due to the regular delivery of milk (which may be on a 24-hour basis), deliveries of packaging and the regular shipment of products.

Noise problems should be taken into consideration when determining plant location. Hazardous wastes consist of oily sludge from gearboxes of moving machines, laboratory waste, cooling agents, oily paper filters, batteries, paint cans etc. At present, in western Europe some of these materials are collected by waste companies. While some waste is incinerated, much is simply dumped.

Water Consumption

As with most food processing operations, water is used extensively for cleaning and sanitising plant and equipment to maintain food hygiene standards. Due to the higher costs of water and effluent disposal that have now been imposed in some countries to reflect environmental costs, considerable reduction in water consumption has been achieved over the past few decades in the dairy processing industry. These improvements are attributed to developments in process control and cleaning practices.

At modern dairy processing plants, a water consumption rate of 1.3–2.5 litres water/kg of milk intake is typical, however 0.8–1.0 litres water/kg of milk intake is possible (Bylund, 1995). To achieve such low consumption requires not only advanced equipment, but also very good housekeeping and awareness among both employees and management.

Effluent Discharge

Dairy processing effluent contains predominantly milk and milk products which have been lost from the process, as well as detergents and acidic and caustic cleaning agents. The constituents present in dairy effluent are milk fat, protein, lactose and lactic acid, as well as sodium, potassium, calcium and chloride. Milk loss to the effluent stream can amount to 0.5–2.5% of the incoming milk, but can be as high as 3–4%.

A major contributing factor to a dairy plant's effluent load is the cumulative effect of minor and, on occasions, major losses of milk. These losses can occur, for example, when pipework is uncoupled during tank transfers or equipment is being rinsed.

The organic pollutant content of dairy effluent is commonly expressed as the 5-day biochemical oxygen demand (BOD5) or as chemical oxygen demand (COD). One litre of whole milk is equivalent to approximately 110,000 mg BOD5 or 210,000 mg COD.

Concentrations of COD in dairy processing effluents vary widely, from 180 to 23,000 mg/L. Low values are associated with milk receipt operations and high values reflect the presence of whey from the production of cheese. A typical COD concentration for effluent from a dairy plant is about 4000 mg/L. This implies that 4% of the milk solids received into the plant is lost to the effluent stream, given that the COD of whole milk is 210,000 mg/L and that effluent COD loads have been estimated to be approximately 8.4 kg/m3 milk intake.

A Danish survey found that effluent loads from dairy processing plants depend, to some extent, on the type of product being produced. The scale of the operation and whether a plant uses batch or continuous processes also have a major influence, particularly for cleaning. This is because small batch processes requires more frequent cleaning. The tendency within the industry towards larger plants is thus favourable in terms of pollutant loading per unit of production.

Water Consumption Survey for Danish Dairy Processing Plants

A survey of 72 Danish dairy companies operating a total of 134 processing plants was conducted in 1989 (Danish EPA, 1991). The product mix of the companies surveyed was as follows: 44 dairies produced butter, 90 produced cheese, 29 were market milk plants and 11 produced concentrates including milk powder. The plants surveyed were all technologically advanced and most claimed that they had reduced the pollutant load of their effluents by 30–50% compared with previous years. The survey found that on average each tonne of milk processed resulted in the production of 1.3 m3 of effluent with the following characteristics:

COD 2000 mg/L

BOD5 1500 mg/L

Fat 150 mg/L

Total nitrogen 100 mg/L

Total phosphorus 30 mg/L

Due to the traditional payment system for raw milk (which is based on the mass or volume delivered plus a separate price or premium for the weight of milk fat), the dairy processing industry has always tried to minimise loss of milk fat. In many countries the payment system now recognises the value of the non-fat milk components. Systems that control the loss of both fat and protein are now common in the industrialised world, but less so in the developing world. The disposal of whey produced during cheese production has always been a major problem in the dairy industry. Whey is the liquid remaining after the recovery of the curds formed by the action of enzymes on milk. It comprises 80–90% of the total volume of milk used in the cheese making process and contains more than half the solids from the original whole milk, including 20% of the protein and most of the lactose. It has a very high organic content, with a COD of approximately 60,000 mg/L (Morr, 1992). Only in the past two decades have technological advances made it economically possible to recover soluble proteins from cheese whey and, to some extent, to recover value from the lactose.

Most dairies are aware that fat and protein losses increase the organic load of the effluent stream and, even in the developing world, the use of grease traps has been common for some decades. Many companies, however, do not take any action to reduce the organic pollution from other milk components. It is becoming more common for dairy companies to be forced by legal or economic pressures to reduce the amount and concentration of pollutants in their effluent streams.

Therefore, at most sites, wastewater treatment or at least pretreatment is necessary to reduce the organic loading to a level that causes minimal environmental damage and does not constitute a health risk. The minimum pretreatment is usually neutralisation of pH, solids sedimentation and fat removal.

Energy Consumption

Energy is used at dairy processing plants for running electric motors on process equipment, for heating, evaporating and drying, for cooling and refrigeration, and for the generation of compressed air.

Approximately 80% of a plant's energy needs is met by the combustion of fossil fuel (gas, oil etc.) to generate steam and hot water for evaporative and heating processes. The remaining 20% or so is met by electricity for running electric motors, refrigeration and lighting.

The energy consumed depends on the range of products being produced. Processes which involve the concentration and drying of milk, whey or buttermilk for example, are very energy intensive. The production of market milk at the other extreme involves only some heat treatment and packaging, and therefore requires considerably less energy.

Energy consumption will also depend on the age and scale of a plant as well as the level of automation. To demonstrate this.

Table Energy Consumption for a Selection of Milk Plants

Type of Plant Total Energy Consumption

Plants producing powdered milk exhibit a wide range of energy efficiencies, depending on the type of evaporation and drying processes that are used. Energy consumption depends on the number of evaporation effects (the number of evaporation units that are used in series) and the efficiency of the powder dryer. Table–6 provides examples of how different evaporation and drying systems can affect the energy efficiency of the process.

Substantial increases in electricity use have resulted from the trend towards automated plant with associated pumping costs and larger evaporators as well as an increase in refrigeration requirements.

High consumption of electricity can also be due to the use of old motors, excessive lighting or possibly a lack of power factor correction.

Table Energy Consumption for Evaporation and Drying Systems

Type of evaporation and drying system Total energy consumption

(GJ/tonne product)

5-effect evaporator and 2-stage drier 13–15

3-effect evaporator and 1-stage drier 22–28

2-effect evaporator and 1-stage drier 40–50.

6

Enzymes in Milk and Cheese Industries

We have been using enzymes for thousands of years to make various drinks and things to eat, such as cheese, yoghurt, bread, beer, and wine. Of course the people who first discovered how to make these knew nothing about enzymes. In this unit we can see how these amazing biological catalysts are used in making these traditional foodstuffs, and also in making new products.

In the field of biotechnology there are many industrial applications that result in biotech products that we use everyday at home. Some of these are food science applications that utilize enzymes to produce or make improvements in the quality of different foods.

In the dairy industry, some enzymes are required for the production of cheeses, yogurt and other dairy products, while others are used in a more specialized fashion to improve texture or flavour. Five of the more common types of enzymes and their role in the dairy industry are described below.

Rennet

Milk contains proteins, specifically caseins, that maintain its liquid form. Proteases are enzymes that are added to milk during cheese production, to hydrolyze caseins, specifically kappa casein, which stabilizes micelle formation preventing coagulation.

Rennet and rennin are general terms for any enzyme used to coagulate milk. Technically rennet is also the term for the lining of a calf's fourth stomach. The most common enzyme isolated from rennet is chymosin. Chymosin can also be obtained from several other animal, microbial or vegetable sources, but indigenous microbial chymosin (from fungi or bacteria) is ineffective for making cheddar and other hard cheeses. Limited supplies of calf rennet have prompted genetic engineering of microbial chymosin by cloning calf prochymosin genes into bacteria.

Bioengineered chymosin may be involved in production of up to 70% of cheese products. While use of bioengineered enzymes spares the lives of calves, it presents ethics issues for those opposed to eating foods prepared with GEMs.

Other Proteases

Milk contains a number of different types of proteins, in addition to the caseins. Cow milk also contains whey proteins such as lactalbumin and lactoglobulin. The denaturing of these whey proteins, using proteases, results in a creamier yogurt product. Destruction of whey proteins is also essential for cheese production.

During production of soft cheeses, whey is separated from the milk after curdling, and may be sold as a nutrient supplement for body building, weight loss, and lowing blood pressure, among other things.

There have even been reports of dietary whey for cancer therapies, and having a role in the induction of insulin production for those with Type 2 diabetes. Proteases are used to produce hydrolyzed whey protein, which is whey protein broken down into shorter polypeptide sequences. Hydrolyzed whey is less likely to cause allergic reactions and is used to prepare supplements for infant formulas and medical uses.

Lactase

Lactase is a glycoside hydrolase enzyme that cuts lactose into it's constituent sugars, galactose and glucose. Without sufficient production of lactase enzyme in the small intestine, humans become lactose intolerant, resulting in discomfort (cramps, gas and diarrhea) in the digestive tract upon ingestion of milk products. Lactase is used commercially to prepare lactose-free products, particularly milk, for such individuals.

It is also used in preparation of ice cream, to make a creamier and sweeter-tasting product. Lactase is usually prepared from *Kluyveromyces* sp. of yeast and *Aspergillus* sp. of fungi.

Catalase

The enzyme Catalase has found limited use in one particular area of cheese production. Hydrogen peroxide is a potent oxidizer and toxic to cells. It is used instead of pasteurization, when making certain cheeses such as Swiss, in order to preserve natural milk enzymes that are beneficial to the end product and flavour development of the cheese.

These enzymes would be destroyed by the high heat of pasteurization. However, residues of hydrogen peroxide in the milk will inhibit the bacterial cultures that are required for the actual cheese production, so all traces of it must be removed. Catalase enzymes are typically obtained from bovine livers or microbial sources, and are added to convert the hydrogen peroxide to water and molecular oxygen.

Lipases

Lipases are used to break down milk fats and give characteristic flavours to cheeses. Stronger flavoured cheeses, for example, the italian cheese, Romano, are prepared using lipases.

The flavour comes from the free fatty acids produced when milk fats are hydrolyzed. Animal lipases are obtained from kid, calf and lamb, while microbial lipase is derived by fermentation with the fungal species *Mucor meihei*. Although microbial lipases are available for cheese-making, they are less specific in what fats they hydrolyze, while the animal enzymes are more partial to short and medium-length fats.

Hydrolysis of the shorter fats is preferred because it results in the desirable taste of many cheeses. Hydrolysis of the longer chain fatty acids can result in either soapiness, or no flavour at all.

Enzymes in Baking Industry

For decades, enzymes such as malt and fungal alpha-amylases have been used in bread-making. Rapid advances in biotechnology have made a number of exciting new enzymes available for the baking industry. The importance of enzymes is likely to increase as consumers demand more natural products free of chemical additives. For example, enzymes can be used to replace potassium bromate, a chemical additive that has been banned in a number of countries.

The dough for bread, rolls, buns and similar products consists of flour, water, yeast, salt and possibly other ingredients such as sugar and fat. Flour consists of gluten, starch, non-starch polysaccharides, lipids and trace amounts of minerals. As soon as the dough is made, the yeast starts to work on the fermentable sugars, transforming them into alcohol and carbon dioxide, which makes the dough rise.

The main component of wheat flour is starch. Amylases can degrade starch and produce small dextrins for the yeast to act upon. There is also a special type of amylase that modifies starch during baking to give a significant anti-staling effect

Table: *Lists some of the bread properties that can be improved using industrial enzymes.*

Enzyme	***Effect***
Amylase	Maximises the fermentation process to obtain an even crumb structure and a high loaf volume
Maltogenic alpha-amylases	Improves shelf-life of bread and cakes
Glucose oxidase	Oxidative reaction with gluten to make weak doughs stronger, drier and more elastic
Lipase	Modifies the natural lipids in flour to strengthen the dough
Lipoxygenase	Bleaching and strengthening dough
Xylanase	Dough conditioning. Easier dough handling and improved crumb structure
Protease	Weakens the gluten to give plastic properties required in doughs for biscuits
Asparginase	Reduces the amount of acrylamid formed during baking

Gluten is a combination of proteins that forms a large network during dough formation. This network holds the gas in during dough proofing and baking. The strength of this gluten network is therefore extremely important for the quality of all bread raised using yeast. Enzymes such as hemicellulases, xylanases, lipases and oxidases can directly or indirectly improve the strength of the gluten network and so improve the quality of the finished bread.

Flour Supplementation

Malt flour and malt extract can be used as enzyme supplements because malt is rich in alpha-amylases. Commercial malt preparations can differ widely in their enzyme activity, whereas an industrial enzyme is supplied with a standardised activity.

The alpha-amylases degrade the damaged starch in wheat flour into small dextrins, which allows yeast to work continuously during dough fermentation, proofing and the early stage of baking. The result is improved bread volume and crumb texture. In addition, the small oligosaccharides and sugars such as glucose and maltose produced by these enzymes enhance the Maillard reactions responsible for the browning of the crust and the development of an attractive baked flavour.

Bread and cake staling is responsible for significant financial loss for both consumers and producers. For instance, every year in the USA bread worth more than USD 1 billion is discarded. However the main saving on prolonging the shelf life is actually savings in transportation and fuel costs due to a more efficient distribution. Staling is associated with a loss of freshness in terms of increased crumb firmness, decreased crumb elasticity and loss of moistness.

Staling is believed to be due to changes in starch structure during storage. When the starch granules revert from a soluble to an insoluble form, they lose their flexibility; the crumb becomes hard and brittle. For decades, emulsifiers have been used as anti-staling agents. However, they actually have a limited anti-staling effect and are subject to special labelling rules.

By contrast, Novozymes' bacterial maltogenic alpha-amylases has been found to have a significant anti-staling effect. It modifies the starch during baking at the temperature when most of the starch starts to gelatinise. The resulting modified starch granules remain more flexible during storage. Bread produced with maltogenic alpha-amylase has a far softer and more elastic crumb than bread produced with distilled monoglycerides as emulsifiers.

Dough Conditioning

Flour contains 2.5-3.5% non-starch polysaccharides, which are large polymers (mainly pentosans) that play an important role in bread quality due to their water absorption capability and interactions with gluten. Although the true mechanism of hemicellulase, pentosanase or xylanase in bread-making has not been clearly demonstrated, it is well known that the addition of certain types of pentosanases or xylanases at the correct dosage can improve dough machinability yielding a more flexible, easier-to-handle dough. Consequently, the dough is more stable and gives better ovenspring during baking, resulting in a larger volume and improved crumb texture.

Normal wheat flour contains 1-1.5% lipids, both polar and non-polar. Some of these lipids, especially the polar lipids such as phosphorlipids and galactolipids are able to stabilise the air bubbles in the gluten matrix. The addition of functional lipases modifies the natural flour lipids so they become better at stabilizing the dough.

This ensures a more stable dough in case of over-fermentation, a larger loaf volume, and significantly improved crumb structure. Because of the more uniform and smaller crumb cells, the crumb texture is silkier and the crumb colour appears to be whiter. It also reduces the need for addition of emulsifiers like DATEM and SSL that otherwise are commonly added to dough in order to stabilise it. This in turn means that emulsifiers can be removed from the label.

Chemical oxidants such as bromates, azodicarbonamide and ascorbic acid have been widely used to strengthen the gluten when making bread. As an alternative, oxidases such as glucose oxidase can partially replace the use of these chemical oxidants and achieve better bread quality.

The Synergistic Effects of Enzymes

Each of the enzymes mentioned above has its own specific substrate in wheat flour dough. For example, lipases work on the lipids, xylanase works on the pentosans, and amylases work on the starch. Because the interaction of these substrates in dough and bread is rather complex, the use of enzyme combinations can have synergistic effects that are not seen if only one enzyme is used, not even at high dosages. Quite often an overdosage of enzymes will have a detrimental effect on either the dough or the bread.

For instance, an overdose of fungal alpha-amylase or hemicellulase/xylanase may result in a dough that is too sticky to be handled by the baker or baking equipment.

It is therefore beneficial for some types of bread formulation to use a combination of lower dosages of alpha-amylase and xylanase with low dosages of lipase or glucose oxidase to achieve optimum dough consistency stability and bread quality. Another example is to use maltogenic alpha-amylase in combination with fungal alpha-amylases and xylanase or lipase to secure optimum crumb softness as well as optimum bread quality in terms of crumb structure, bread volume, etc.

Reduction of Acrylamide Content in Food Products

During recent years it has been shown that the amount of the potentially carcinogenic substance acrylamide is relatively high in a number of cereal and potato based products like biscuits, crackers, crisp bread, French fries and potato crisps. This is a substance that is formed at high temperatures when the amino acid asparagine reacts with a reducing sugar like glucose. To meet this issue the enzyme asparaginase have been developed in order to reduce the formation of acrylamide. Asparaginase converts asparagine to aspartic acid which does not take part in the formation of acrylamide. The use of an asparginase is able to reduce the formation of acrylamide with up to 90%.

Enzymes in Fruit Juices and Beverages Industry

Enzyme Used in Fruit Juice Manufacturing

Enzymes are, widely used in fruit Juice and brewing industries to achieve specific objectives, which can not be otherwise, achieved. In addition, brewing industry employs varied but specific strains of yeast, e.g., S. cerevisiae, S. carisbergensis, etc., for, fermentation of the brew.

The cloudiness of fruit juices and wines is mainly due to pectins, which exhibit are usually associated with other plant polymers and even cell debris. The pectins are digested by pecteolytic enzymes prepared from A. niger. These enzyme preparations are mixtures of the following:

(i) polygalacturonase (random digestion of pectin),

(ii) pectin esterase (removal of methyl esters and release of methanol; pectin become degradable by polygalacturonase),

(iii) pectin lyase (cleaves pectins into oligosaccharides even without pectin esterase action), and

(iv) hemicellulase (a mixture of enzymes degrading hemicellulose). They are generally added directly into the fruit pulp at about 20 U/l.

The four enzymes present in the mixture act synergistically to accomplish, a task, which can not be achieved by mechanical means. Treatment of fruit pulp with pecteolytic enzyme mixture gives the following benefits:

(i) Elimination of juice/wine cloudiness,

(ii) reduced solution viscosity,

(iii) Increased juice yields, e.g., a 15% increase in case of white grapes, and

(iv) Shorter fermentation period in case of wine making. In addition,

(v) pectins stabilize the cell debris in a colloidal state; but once pectins are digested, the debris precipitate and are removed by filtration.

Uses of Enzymes in Fruit Juice Wine Brewing Distilling Industries

One of the major problems in the preparation of fruit juices and wine is cloudiness due primarily to the presence of pectins. These consist primarily of a-1, 4-anhydrogalacturonic acid polymers, with varying degrees of methyl esterification. They are associated with other plant polymers and, after homogenization, with the cell debris. The cloudiness that they cause is difficult to remove except by enzymic hydrolysis. Such treatment also has the additional benefits of reducing the solution viscosity, increasing the volume of juice produce (e.g. the yield of juice from white grapes can be raised by 15%), subtle but generally beneficial changes in the flavour and, in the case of wine-making, shorter fermentation times.

The enzymes used in brewing are needed for saccharification of starch (bacterial and fungal a-amylases), breakdown of barley a-1.4- and a-1,3-linked glucan (b-glucanase) and hydrolysis of protein (neutral protease) t increase the (later) fermentation rate, particularly in the production of high-gravity beer, where extra protein is added. Cellulases are also occasionally used, particularly where wheat is used as adjunct to help break down the barley is b-glucans.

Due to the extreme heat stability of the B. amyloliquefaciens a-amylase where this is used, the wort must be boiled for a much longer period (e.g. 30 minutes) to inactivate it prior to fermentation. Papain is used in the later post-fermentation stages of beer-making to prevent the occurrence of protein-and tannin- containing 'chill-haze' otherwise formed on cooling the beer.

Recently, 'light' beers, of lower calorific content, have become more popular. These require a higher degree of saccharification at lower starch concentrations to reduce the alcohol and total solid content of the beer. This may be achieved by the use of glucoamylase and/or fungal a-amylase during the fermentation.

A new enzyme preparation of fungal pectin lyase (EC 4.2.2.10) was shown to be useful for the production of cranberry juice and clarification of apple juice in the food industry (Semenova et al., 2006). A comparative study showed that the preparation of pectin lyase is competitive with commercial pectinase products. The molecular weight of homogeneous pectin lyase was 38 kDa. Properties of the homogenous enzyme were studies. This enzyme was most efficient in removing highly esterified pectin.

The zygomycete microfungus R. microsporus var. microsporus produced a 1,3-1,4-beta-D-glucan 4-glucanhydrolase (EC 3.21.73) which was able to hydrolyse beta-D-glucan that contains both the 1,3-and 1,4-bonds (barley beta-glucans) (Celestino et al., 2006). Its molecular mass was 33.7 kDa.

Maximum activity was detected at pH values in the range of 4-5, and temperatures in the range of 50-60°C. The enzyme was able to reduce both the viscosity of the brewermash and the filtration time, indicating its potential value for the brewing industry Landbo et al, (2006) examined the clarification and haze-diminishing effects of alternative clarification strategies on black current juice including centrifugation and addition of acidic protease and pectinolytic enzyme preparation and gallic acid.

The extent of clarification and haze diminishment varied after individual treatment with five different acidic proteases, but one of the protease preparations, Enzeco, derived from Aspergillus niger, consistently tended to perform best.

The individual and interactive effects on juice turbidity, total phenols, and total anthocyanin content of clarification treatment involving the use of two selected acid proteases (Enzeco and Novozyme 89L), a pectinase (Pectinex BE 3-L), and gallic acid were evaluated in a full factorial 2(4) experimental design. Haze development during cold storage decreases when gallic acid or any of the enzyme preparation was employed individually, but negative interaction effects resulted when the pectinase was employed in combination with any of the proteases.

After 28 storage days at 2°C, the lowest levels of haze formation were achieved when the Enzeco protease preparation, added at 0.025 g/L, was added with 0.050 g/L of gallic acid and allowed to react in the juice for 90 min. at 50°C. The corresponding anthocyanin reduction was approximately 12% (compared to approximately 30% with gelatin silica soltreatment).

The data support the hypothesis that phenol-protein interactions are involved in juice turbidity development during cold storage of berry juices and demonstrate that pre-centrifugation and protease-assisted clarification show promise as an alternative, phenolic-retaining clarification strategy in black currant juice processing.

At monosodium glutamate (MSG) biosensor with immobilized L-glutamate oxidase (L-GLOD) has been developed and studied for analysis of MSG in sauces, soups, etc. (Basu et al., 2006). The immobilized enzymatic membrane was attached with oxygen electrode with a push-cap system.

The detection limit of the sensor was 1 mg/dl and the standard curve was found to be linear up to 20 mg/dl. Response time of the sensor was 2 minutes. Cross-linking with glutaraldehyde in the presence of bovine serum albumin (BSA) as a spacer molecule has been used for immobilization.

Optimization of the sensor was done with an increase in L-GOLD concentration (6.3-31.5 IU)(and also with increase in loading volume of enzyme solution (5-20 microL). Optimization of pH and temperature was also studied. The permeability of O2 through different membrane was studied with and without immobolized L-GLOD. The enzymatic membrane was used for 20 measurements and stability of membrane was observed.

Enzymes in Meat Industries

Papain

Papain, also known as papaya proteinase I, is a cysteine protease (EC 3.4.22.2) enzyme present in papaya (*Carica papaya*) and mountain papaya (*Vasconcellea cundinamarcensis*).

Papain Family

Papain belongs to a family of related proteins with a wide variety of activities, including endopeptidases, aminopeptidases, dipeptidyl peptidases and enzymes with both exo- and endo-peptidase activity. Members of the papain family are widespread, found in baculovirus, eubacteria, yeast, and practically all protozoa, plants and mammals.

The proteins are typically lysosomal or secreted, and proteolytic cleavage of the propeptide is required for enzyme activation, although bleomycin hydrolase is cytosolic in fungi and mammals.

Papain-like cysteine proteinases are essentially synthesised as inactive proenzymes (zymogens) with N-terminal propeptide regions. The activation process of these enzymes includes the removal of propeptide regions. The propeptide regions serve a variety of functions in vivo and in vitro.

The pro-region is required for the proper folding of the newly synthesised enzyme, the inactivation of the peptidase domain and stabilisation of the enzyme against denaturing at neutral to alkaline pH conditions. Amino acid residues within the pro-region mediate their membrane association, and play a role in the transport of the proenzyme to lysosomes. Among the most notable features of propeptides is their ability to inhibit the activity of their cognate enzymes and that certain propeptides exhibit high selectivity for inhibition of the peptidases from which they originate.

Structure

The papain precursor protein contains 345aa, and consists of a signal sequence (1-18), a propeptide (19-133) and the mature peptide (134-345). The amino acid numbers are based on the mature peptide. The protein is stabilised by 3 disulfide bridges. Its 3D structure consists of 2 distinct structural domains with a cleft between them. This cleft contains the active site, which contains a catalytic diad that has been likened to the catalytic triad of chymotrypsin. The catalytic diad is made up of the amino acids - cysteine-25 (from which it gets its classification) and histidine-159. Aspartate-158 was thought to play a role analogous to the role of aspartate in the serine protease catalytic triad, but that has since then been disproved. None-the-less, it does appear to play a role.

Function

The mechanism by which it breaks peptide bonds involves deprotonation of Cys-25 by His-159. Asparagine-175 helps to orient the imidazole ring of His-159 to allow this deprotonation to take place. Cys-25 then performs a nucleophilic attack on the carbonyl carbon of a peptide backbone. This frees the amino terminal of the peptide, and forms a covalent acyl-enzyme intermediate. The enzyme is then deacylated by a water molecule, and releases the carboxy terminal portion of the peptide. In immunology, papain is known to cleave the Fc (crystallisable) portion of immunoglobulins (antibodies) from the Fab (antigen-binding) portion.

Uses

Its utility is in breaking down tough meat fibers and has been utilized for thousands of years in its native South America. It is sold as a component in powdered meat tenderizer available in most supermarkets. Papain, in the form of a meat tenderizer such as Adolph's, made into a paste with water, is also a home remedy treatment for jellyfish, bee, and yellow jacket (wasps) stings; mosquito bites; and possibly stingray wounds, breaking down the protein toxins in the venom.

Papain can be used to dissociate cells in the first step of cell culture preparations. A ten-minute treatment of small tissue pieces (less than 1 mm cubed) will allow papain to begin breaking down the extracellular matrix molecules holding the cells together. After ten minutes, the tissue should be treated with a protease inhibitor solution to stop the protease action (if left untreated, papain's activity will lead to complete lysis of the cells).

The tissue must then be triturated (passed quickly up and down through a Pasteur pipette) in order to break up the pieces of tissue into a single cell suspension.

It is also used as an ingredient in various enzymatic debriding preparations, notably Accuzyme. These are used in the care of some chronic wounds to clean up dead tissue.

Papain can also be found as an ingredient in some toothpastes or mints as teeth-whitener. Its whitening effect in toothpastes and mints is minimal, however, because the papain is present in low concentrations, and is quickly diluted by saliva. It would take several months of using the whitening product to have noticeably whiter teeth.

It is the main ingredient of Papacarie, a gel used for chemomechanical dental caries removal. Besides the advantage of avoiding the use of rotary cutting tools, it does not interfere in the bond strength of restorative materials to dentin.

Immunoglobulins

An antibody digested by papain yields three fragments: two 50 kDa Fab fragments and one 50kDa Fc fragment. The papain-digested antibody is unable to promote agglutination, precipitation, opsonization, and lysis.

Bromelain

Bromelain can refer to one of two protease enzymes extracted from the plant family Bromeliaceae, or it can refer to a combination of those enzymes along with other compounds produced in an extract.

Extract Components

Bromelain extract is a mixture of protein-digesting enzymes—called proteolytic enzymes or proteases and several other substances in smaller quantities. The proteolytic enzymes are referred to as sulfhydryl proteases since a free sulfhydryl group of a cysteine side-chain is required for function. The two main enzymes are:

- Stem bromelain - EC 3.4.22.32
- Fruit bromelain - EC 3.4.22.33

The other substances typically include peroxidase, acid phosphatase, protease inhibitors, and calcium.

History

The first isolation of bromelain was recorded by the Venezuelan chemist Vicente Marcano (BU1 1.Phar. 5,77) in 1891 from the fruit of pineapple. In 1892, Chittenden, assisted by Joslin and Meara, investigated the matter fully (Trans. Conn. Acad. Arts Sci. 8, 281-308), and called it 'bromelin'. Later the term 'bromelain' was introduced and originally applied to any protease from any plant member of the plant family Bromeliaceae.

Bromelain was first introduced as a therapeutic supplement in 1957. Research on bromelain apparently was first conducted in Hawaii but more recently has been conducted in countries in Asia, Europe and Latin America. Germany has recently taken a great interest in bromelain research; bromelain is currently the 13th most widely used herbal medicine in Germany.

Source

Bromelain is present in all parts of the pineapple plant (*Ananas sp.*), but the stem is the most common commercial source, presumably because it is readily available after the fruit has been harvested. Pineapples have had a long tradition as a medicinal plant among the natives of South and Central America.

Uses

Meat Tenderizing

Along with papain, bromelain is one of the most popular substances to use for meat tenderizing.

Today, about 90% of meat tenderizer is used in consumer households. Bromelain is sold in a powdered form, which is combined with a marinade, or directly sprinkled on the uncooked meat. The enzyme will penetrate the meat, and by a process called forking, cause the meat to be tender and palatable when cooked. If the enzyme is allowed to work for too long, the meat may become too "mushy" for many consumers' preferences. Cooked or canned pineapple does not have a tenderizing effect, as the enzymes are heat labile.

Some prepared meat products, such as meatballs, and commercially available marinades include pineapple and/or pineapple-derived ingredients.

Apart from the mushiness of meat that has been over-tenderized, the activity of bromelain and similarly proteolytic plant enzymes can be undesirable where it is inappropriate. In dishes that depend on their protein content for important attributes, uncooked pineapple or its juice can be a nuisance. For example, some dishes such as brawn and jelly [called gelatin dessert (Jell-o) in the US] rely on the setting of gelatin. They will not set if they contain raw pineapple or pineapple juice. Raw figs, papaya, and similarly proteolytic vegetable matter causes similar problems.

To prevent the problem, the fruit should first be cooked, or at least parboiled sufficiently to inactivate the enzymes. Thorough heating to above about 65 degrees C will generally suffice. This problem does not affect dishes based on nonproteinaceaous gelling agents, such as agar, though such agents may be prevented from gelling by too much heating in acid recipes.

7

Technological Principles of Pasteurization

Physical and chemical factors which influence pasteurization process are the following:

a. temperature and time;

b. acidity of the products;

c. air remaining in containers.

Pasteurization processes. In pasteurising certain acid juices for example, there are two categories of processes:

Low pasteurization where pasteurization time is in the order of minutes and related to the temperature used; two typical temperature/time combinations are as following:

63° C to 65° C over 30 minutes or

75° C over 8 to 10 minutes.

Pasteurization temperature and time will vary according to:

- nature of product; initial degree of contamination;
- pasteurized product storage conditions and shelf life required.

In this first category of pasteurization processes it is possible to define three phases:

- heating to a fixed temperature;
- maintaining this temperature over the established time period (= pasteurization time);
- cooling the pasteurized products: natural (slow) or forced cooling.

Rapid, high or flash pasteurization is characterized by a pasteurization time in the order of seconds and temperatures of about 85° to 90° C or more, depending on holding time. Typical temperature/time combinations are as follows:

88° C (190° F) for 1 minute;

100° C for 12 seconds;

121°C for 2 seconds.

While bacterial destruction is very nearly equivalent in low and in high pasteurization processes, the 121° C/2 seconds treatment give the best quality products in respect of flavour and vitamin retention. Such short holding times, however, require special equipment which is more difficult to design and generally is more expensive than the 63-65 ° C/30 minutes type of processing equipment.

In flash pasteurization the product is heated up rapidly to pasteurization temperature, maintained at this temperature for the required time, then rapidly cooled down to the temperature for filling, which will be performed in aseptic conditions in sterile receptacles. Taking into account the short time and rapid performance of this operation, flash pasteurization can only be achieved in continuous process, using heat exchangers.

Industrial applications of pasteurization process are mainly used as a means of preservation for fruits and vegetable juices and specially for tomato juice.

Thermopenetration. The thermopenetration problem is extremely important, especially in the case of the pasteurization of products packed in glass containers because it is the determining factor for the success of the whole operation.

During pasteurization it is necessary that a sufficient heat quantity is transferred through the receptacle walls; this is in order that the product temperature rises sufficiently to be lethal to micro-organisms throughout the product mass.

The most suitable and practical method to speed up thermopenetration is the movement of receptacles during the pasteurization process. Rapid rotation of receptacles around their axis is an efficient means to accelerate heat transfer, because this has the effect, among others of rapidly mixing the contents.

The critical speed of for this movement is generally about 70 rotations per minute (RPM). This enables a more uniform heating of products, reducing heating time and organoleptic degradation.

Heating may precede or follow packaging. These principles of different temperature time combinations very largely determine the design parameters for heat preservation equipment and commercial practices. The food processor will employ no less than that heat treatment which gives the necessary degree of micro-organism destruction. This is further ensured by periodic inspection by local sanitary authorities or by the importing countries sanitary services. However, the food processor also will want to use the mildest effective heat treatment to ensure highest food quality.

It is convenient to separate heat preservation practices into two broad categories: one involves heating of foods in their final containers, the other employs heat prior to packaging. The latter category includes methods that are inherently less damaging to food quality, where the food can be readily subdivided (such as liquids) for rapid heat exchange. However, these methods then require packaging under aseptic or nearly aseptic conditions to prevent or at least minimise recontamination.

On the other hand, heating within the package frequently is less costly and produces quite acceptable quality with the majority of foods and most of our present canned food supply is heated in the package.

In practice, therefore, most of the canned food produced locally in developing countries should be heated within the package.

Food Irradiation

Food irradiation is one of the food processing technologies available to the food industry to control organisms that cause food-borne diseases and to reduce food losses due to spoilage and deterioration. Food irradiation technology offers some advantages over conventional processes. Each application should be evaluated on its own merit as to whether irradiation provides a technical and economical solution that is better than traditional processing methods.

Table: *Possible causes of spoilage (real or apparent) in canned goods*

Condition of can	*Action to be taken to identify cause*
Insufficient vacuum or headspace	Check vacuum and headspace in relation to storage temperature and altitude
"Springer" or "flipper"	Cool can to 15°C and check if still domed. Check can for denting, if possible measure headspace volume change brought about by dents, by comparing can volume with volume of a sound can
Hydrogen swell	Check degree of detinning in can especially at the liquid level. Look for scratches or pinholes in lacquer or tin coating. Check if can is still domed on cooling to 15°C.
"Hard" or "soft swell"	Leaker spoilage. Check can for gross seam faults, perforation due to corrosion or damage to seams. Examine contents for signs of spoilage and can interior for detinning at air/product interface.

Source: FAO/WFP, 1970

Applications

For products where irradiation is permitted, commercial applications depend on a number of factors including the demand for the benefits provided, competitiveness with alternative processes and the willingness of consumers to buy irradiated food products. There are a number of applications of food irradiation. For each application it is important to determine the optimum dosage range required to achieve the desired effect. Too high a dosage can produce undesirable changes in texture, colour and taste of foods.

Shelf-life extension. Irradiation can extend the shelf-life of foods in a number of ways. By reducing the number of spoilage organisms (bacteria, mould, fungi), irradiation can lengthen the shelf life of fruits and vegetables.

Since ionising radiation interferes with cell division, it can be used as an alternative to chemicals to inhibit sprouting and thereby extend the shelf life of potatoes, onions and garlic. Exposure of fruits and vegetables to ionising radiation slows their rate of ripening. Strawberries, for example, have been found to be suitable for irradiation. Their shelf-life can be extended three-fold, from 5 to 15 days.

Disinfestation. Ionising radiation can also be used as an alternative to chemical fumigants for disinfestation of grains, spices, fruits and vegetables. Many countries prohibit the importation of products suspected of being contaminated with live insects to protect the importing country's agricultural base. With the banning of certain chemical fumigants, irradiation has the potential to facilitate the international shipment of food products.

Global Developments

Consensus on Wholesomeness

In 1980, an FAD/IAEA/WHO Expert Committee reviewed in detail all the accumulated data on food irradiation from the past 40 years.

The Expert Committee concluded that irradiation to an overall dose of 10 kGy (kilograys) presents no toxicological hazard and introduces no special nutritional or microbiological problems, thus establishing the wholesomeness of irradiated foods up to an overall average absorbed dose of 10 kGy.

Data were insufficient to formulate conclusions on applications of food irradiation above 10 kGy. Data on radiation chemistry, nutritional and microbiological aspects of food treated above 10 kGy is currently being compiled.

In 1983, the Codex Alimentarius Commission, an international group that develops global food standards for the FAO and the WHO, incorporated the 1980 Expert Committee's conclusions regarding the wholesomeness of irradiated foods into the Codex General Standard for Irradiated Foods. This proposed international standard was submitted to member countries to accept or to modify according to individual country needs. Currently most countries that allow food irradiation approve its use on a case-by-case basis.

The Codex Alimentarius Commission has also adopted a Recommended International Code of Practice for the Operation of Radiation Facilities for the Treatment of Foods. It is intended to serve as a guide for irradiator operators and government regulators.

International Trade

More than 30 countries have given clearances for the use of food irradiation to process some 40 food items and approximately 30 facilities world-wide treat food by irradiation processing. Approvals for additional items are being considered in many countries and many food irradiation facilities are being planned. It was anticipated in 1988 that by 1990 there could be approximately 50 commercial/demonstration irradiators in 25 countries. Table below shows commercial applications of food irradiation to fruits and vegetables by country.

Table: *International commercial applications of radiation for fruit and vegetables*

Country	*Location (application date)*	*Food Commodity*
Argentina	Buenos Aires (1986)	Spinach
Belgium	Fleurus (1981)	Dehydrated vegetables
Brazil	Sao Paulo (1985)	Dehydrated vegetables
Chile	Santiago (1983)	Dehydrated vegetables onions, potatoes
China	Shanghai (1985	Potatoes
Cuba	Havana (1987)	Potatoes, onions
German Dem. Rep	Weideroda (1983)	Onions, garlic
	Spickendorf (1986)	Onions
Japan	Hokkaido (1973)	Potatoes
Korea	Seoul (1985	Garlic powder
Netherlands	Ede (1978)	Dehydrated vegetables
South Africa	Johannesburg (1981)	Dehydrated vegetables
	Tzaneen (1981)	Fruits, onions, potatoes
Thailand	Bangkok (1971)	Onions

Auxiliary Raw Materials

Auxiliary raw materials used in fruit and vegetable processing technologies play a major role in the determination of their physical and chemical characteristics, sensory properties and nutritive value.

Water

Water is one of the essential factors in the activity of the processing centres; according to the final utilisation, water can be classified in three categories:

a. for technological utilisation (when it comes into direct contact with raw materials and enters in the finished product's composition),

b. for steam generators and

c. for receptacle cooling, washing of equipment and general hygiene.

Water for Technological Uses

Water coming into direct contact with the raw materials used for processing (washing, blanching, etc.) or that used as filling liquid of some canned products, must be of drinking water quality in terms of its physico-chemical and microbiological conditions.

More important even than fulfilling drinking water standards, water used for these purposes pests certain specific characteristics related to the technological step or the raw material treated during the processing.

When very hard water is used for blanching vegetables some pecto-calcium and pectomagnesium complexes are formed which starts the hardening of vegetable tissues. This process continues over the pasteurization of the finished product.

When fruit is processed in sugar syrup, the use of hard water for the syrup preparation could induce the formation of a pectin-sugar-acid gel facilitated by the medium pH and presence of calcium salts.

Soft water has negative consequences associated with mineral and hydrosoluble substances and losses during blanching of vegetables.

For some specific products such as peeled tomatoes, green beans and fine texture fruit, the addition of calcium salts (chlorure, sulphate, etc.) is employed to correct for low texture.

The water hardness is an essential factor when used as filling liquid for canned products; ideally, the hardness of the water should be adapted to the raw material species used for canning. Thus a hardness of 3° is good for beans, 5° to 9° for green peas, green beans, and for fruit and vegetables with a tendency to disintegrate should use even harder water.

In the technological process of cucumber and gherkin preservation by natural acidification (lactic fermentation), water hardness has a paramount role. The literature maintains that as far as texture is concerned, the best results are obtained by using very hard water (about 30°); but since high magnesium and salt content has a negative effect on taste, in practice it is recommended that a water hardness of about 10° be used, which gives satisfactory results from both points of view.

Oxygen present in water can act as a corrosion factor in metal receptacles but this negative influence can be eliminated by preliminary boiling.

An important factor is pH. Water for canning must be neutral or slightly basic. Acid water plays a major role in corrosion which is evident both on receptacles and on iron or copper equipment, where changes of product colour will be induced. More dangerous is the attack on lead pipes or to the mix used for can sticking and this can render the product toxic. For these reasons an acid water must be neutralised before use.

Water for Steam Generators

Two main conditions must be fulfilled:

- hardness has to be as low as possible, even zero, because precipitation of calcium salts can lead to the formation of encrustations ("crusts") in pipes and on equipment walls. For this reason, water treatment is practiced in the majority of installations;
- from a bacteriological point of view, the iron-bacteria must be eliminated with biological filters or oxidising substances. This is necessary in order to avoid iron hydroxide formed during equipment exploitation to deposit on the inner walls of the pipes.

The elimination of iron-bacteria is also of importance for the water used in processing steps.

Water for Receptacle Cooling and General Hygiene

This should be of drinking water standard. Where this is not available in sufficient quantity, the use of industrial water is acceptable but only for cleaning of production rooms/workshops.

Sweeteners

Sugar

Sugar is the conventional name applied to sucrose. Physically there is icing, granulated and lump sugar. In fruit and vegetable processing, sugar is used only in its granulated form; this quality must be in the form of uniform crystals, white, shining and completely soluble in water.

Concentration of various sugar solutions can be rapidly measured by refractometer reading or with areometers graduated in various ways: Brix, Baumé, etc.

Sugar solubility in water is dependent upon temperature; for example, in order to obtain a saturated solution, one must dissolve 2040 g in one litre of water at 20° C and 4870 g at 100° C. Taking into account this temperature related solubility, in practice the majority of sugar solutions are prepared by heating the water. Water should be as soft as possible because the calcium salts can precipitate on boiling.

Corn Syrup (Liquid Glucose)

Corn syrup is obtained industrially by acid or enzymatic starch hydrolysis, using as starting raw materials maize (corn) or potatoes. In fruit processing, mainly in the production of marmalades, it is possible to use corn syrup. The average composition of this corn syrup is of about 32-40% dextrose (glucose), about 40% dextrins and 18-20% moisture. Sweetening power is 50% compared with sucrose.

In a 10%-20% proportion with sucrose, addition of corn syrup has certain advantages:

a. it improves the shine and texture of marmalade;

b. it prevents "sugaring" defect and

c. it reduces the too sweet taste of finished products obtained with sugar alone.

Salt

Salt is used in order to give to the finished products a specifically salty taste and as a preserving substance. From a chemical point of view the term salt means sodium chloride but in practice the product is never in a pure state. The presence of a significant quantity of magnesium chloride increases the hygroscopicity, gives a bitter taste and can induce corrosion of receptacles.

Food Acids

Acetic acid is in use as solutions of various concentrations which are known under the generic name of vinegar. Vinegar can be obtained:

a) from wine, alcohol, cider, beer, etc. by fermentation;

b) by dilution of acetic acid obtained by dry wood distillation or by synthesis.

From a quality point of view, wine vinegar is preferred, as it has a more pleasant flavour. In order to improve taste, other vinegar types are usually flavoured with spices.

In addition to its spicing and flavouring role, vinegar is used and acts as a preservation agent for some vegetables: cucumbers, acidified vegetables, etc.

- Citric acid
- Tartric acid
- Lactic acid.

Pectic Preparations

In fruit processing there many preparations and mixes known as "pectin" are used as liquid or powder extracts.

From a practical point of view, the pectic preparations are classified as:

- strong pectins obtained from apples or citrus fruit peel; this category gives gels rich in sugar;
- weak pectins which gives gels with low proportion of sugar or even without sugar but with the addition of calcium salts.

The fruit industry uses mainly strong pectins. These preparations are characterised, from a commercial point of view, by the capacity of gelification, expressed in degrees. The degree of gelification represents the quantity of sugar in grams able to be transformed in a standard gel (65% sugar and pH=3) by 1 g pectin.

Intensive Sweeteners

"Calorie-reduced" and "low-calorie foods are widely used and are cornering an increasing share of the market. Sweeteners are making an important contribution to the manufacture of sweet foods in these categories. They make it possible to manufacture sweet products without "sweet" being necessarily synonymous with "high-calorie".

Diabetics need to restrict their intake of sugar and various carbohydrates similar to sugar or avoid them altogether. Sweeteners enable diabetics to enjoy sweet tastes without changing their lifestyle. Sweeteners do not contribute to the development of tooth decay; they do not degrade in the mouth to form the acids which are responsible for caries. Thus sweeteners offer consumers a number of advantageous and favourable properties above and beyond merely reducing calories.

Sunett is the trade mark of Hoechst AG for its high intensity sweetener acesulfame K. As an ingredient, it can be used for sweetening all foods produced industrially or at home, or to produce tabletop sweeteners. Like all other sweeteners, Sunett tastes more intensely sweet than sugar; it is about 200 times sweeter than sugar compared with a dilute stock solution.

Synergism. - Sunett is notable for its pronounced synergism with other sweeteners. The synergistic effect leads as quantitative synergism to an intensification of the overall sweetness and as qualitative synergism to an improved taste. The synergism results in a marked intensification of the sweet taste of the blends, which can amount to 30-50% at usual concentrations.

The favourable properties of Sunett, particularly its synergistic behaviour, can be used advantageously in the Sunett-Multi-Sweetener Concept. For high sweetness levels, blends of Sunett with other sweeteners, e.g. aspartame, are particularly favourable in many applications. The synergistic effect of the blend results in a taste that is particularly pleasant and rounded.

Alone or in blends with other sweeteners, Sunett is used mainly where only sweetness is important and no other properties of the foods are to be affected. This applies particularly to beverages (fruit syrups and juices, carbonated beverages, etc.).

In a number of food products, sweet carbohydrates do not only provide the sweet taste; they have other functions to fulfil; for example, they act as bulking or texturing agents and as preservatives by reducing the water activity. In these types of products, Sunett and other sweeteners cannot be used on their own. They must be combined with other substances which perform the required functions. These may be bulking agents or sugar substitutes (for example polyols: sorbitol, mannitol, xylitol). Sunett can be combined with both groups of substances.

Uses. In carbonated soft drinks blends of Sunett with other sweet-tasting substances are often recommended. Beverages formulated using the Sunett Multi-Sweetener Concept, i.e. blends of Sunett with other sweeteners, are usually preferred because of their particularly well balanced and rounded sweetness profile. The taste of these blends is often superiors to single sweeteners.

Sunett is compatible with sugar and other sweet-tasting carbohydrates, both technically and in terms of taste. It can therefore be used in the production of soft drinks with a reduced sugar content. Drinks based on fructose and Sunett are suitable for diabetics. Soft drinks with a reduced sugar content show an excellent taste quality when Sunett or Sunett blends are added to bring them up to the usual level of sweetness.

Fruit nectars and fruit juice drinks differ from most carbonated drinks in that they contain a noticeable amount of sweet carbohydrates provided by the fruit juice. The amount of carbohydrate may vary depending on the type of fruit and amount of juice used.

Amounts of up to 200 mg/l Sunett are often adequate as a single sweetener for the popular types of fruit nectars. With blends of Sunett and aspartame quantities in the order of 100150 mg/l Sunett and 50 mg/l aspartame or 60-70 mg/l of both Sunett and aspartame are sufficient.

Sunett offers such excellent stability that end products, such as drinks, show no reduction in sweetness performance during normal processing methods and storage periods. Sunett can withstand pasteurization, hot filling and aseptic filling without any loss of sweetness.

Jams and marmalades. Sugar contributes a great deal to the texture and stability of conventional jams and marmalades. For the production of sugar-free products either the sugar must be replaced by comparable amounts of sugar substitutes or some of the functions of the sugar must be taken only by other components, such as suitable gelling agents.

Sugar-free jams and marmalades containing sweeteners are more susceptible to microorganisms than sugar-containing products. The risk of spoilage due to yeast fermentation or moulding can be prevented by pasteurization. However, this is only feasible for small jars which will be quickly consumed once they are opened.

In all other cases, it is advisable to add 0.05-0.1% potassium sorbate as a preservative, wherever this is permitted under the relevant food regulations.

For sugar-free jams and marmalades, concentrations in the range of 500-2000 mg Sunett/kg of the finished product are appropriate. It is advisable to add Sunett in the form of an aqueous stock solution towards the end of the boiling process. Care must be taken to ensure that Sunett is evenly dispersed throughout the whole batch. Because of the excellent compatibility of Sunett with sugar alcohols, fruit jams and marmalades using these ingredients offer an outstanding taste.

Fruit preserves. Sunett can be used for the production of sugar-free or sugar-reduced fruit preserves. At the normal pH values for fruit preserves Sunett can be added even before pasteurization, as the sweetness is not impaired under the usual thermal treatment conditions. Sunett also withstands the usual storage periods without loss of sweetness.

NutraSweet (r) is the commercial name of aspartame (APM), a new sweetener from G.D. SEARLE & Co. which can be used in the most foods in order to give the same taste as sugar. NutraSweet is about 180 to 200 times sweeter than sucrose (sugar) and this value depends on pH, temperature and the type of flavour.

NutraSweet can be used as mentioned above in a mix with Sunett or alone in all sugar-free or calorie-reduced fruit jams, marmalades and preserves. Like other sweeteners, NutraSweet does not promote tooth decay.

International regulatory status. Both Sunett and NutraSweet are widely accepted by food laws in the majority of countries. The following are the main fruit products where Sunett is an accepted sweetener:

- low-joule prepared jelly: max. 500 mg/kg;
- canned fruit without added sugar: max. 500 mg/kg;
- beverages including calorie-reduced fruit nectars: max. 600 mg/l;
- calorie-reduced jams;

- calorie-reduced and dietetic fruit compotes;
- canned fruit and vegetables, fruit puree, jams and marmalades;
- juices, nectars and juice based beverages;
- jams, marmalades and related products: max. 300 mg/kg.

Packaging Materials

Requirements and Functions of Food Containers

The following are among the more important general requirements and functions of food packaging materials/containers:

a. they must be non-toxic and compatible with the specific foods;
b. sanitary protection;
c. moisture and fat protection;
d. gas and odour protection;
e. light protection;
f. resistance to impact;
g. transparency;
h. tamperproofness;
i. ease of opening;
j. pouring features;
k. reseal features;
l. ease of disposal;
m. size, shape, weight limitations;
n. appearance, printability;
o. low cost;
p. special features.

Primary and Secondary Containers

The terms primary and secondary containers have been used. Some foods are provided with efficient primary containers by nature, such as nuts, oranges, eggs and the like. In packaging these, we generally need only a secondary outer box, wrap, or drum to hold units together and give gross protection.

Other foods such as milk, dried eggs and fruit concentrates often will be filled into primary containers such as plastic liners which are then packaged within protective cartons or drums. In this case the secondary container provided by the carton or drum greatly minimises the requirements that must be met by the primary container.

Except in special instances, secondary containers are not designed to be highly impervious to water vapour and other gases, especially at zones of sealing, dependence for this being placed upon the primary container.

Since primary containers by definition are those which come in direct contact with the food, we will be far more concerned with them than with secondary containers.

Hermetic Closure

Two conditions of the greatest significance in packaging are hermetic and non-hermetic closure.

The term hermetic means a container which is absolutely impermeable to gases and vapours throughout its entirety, including its seams.

Such a container, as long as it remains intact, will automatically be impervious to bacteria, yeasts, moulds, and dirt from dust and other sources since all of these agents are considerably larger than gas or water vapour molecules.

On the other hand, a container which prevents entry of micro-organisms, in many instances will be non-hermetic. A container that is hermetic not only will protect the product from moisture gain or loss, and from oxygen pick-up from the atmosphere, but is essential for strict vacuum and pressure packaging.

The most common hermetic containers are rigid metal cans and glass bottles, although faulty closures can make them non-hermetic. With very rare exceptions flexible packages are not truly hermetic for one or more of the following reasons.

First, the thin flexible films, even when they do not contain minute pinholes, generally are not completely gas and water-vapour impermeable although the rates of gas and water vapour transfer may be exceptionally slow; second, the seals are generally good but imperfect; and third, even where film materials may be gas- and water-vapour-tight, such as certain gages of aluminium foil, flexing of packages and pouches leads to minute pinholes and crease holes.

Hermetic rigid aluminium containers can be readily formed without side seams or bottom end seams. The only seam then to make hermetic is the top end double seam, which may be closed on regular tin can sealing equipment.

Glass containers are hermetic provided the lids are tight. Lids will have inside rings of plastic or cork. Many glass containers are vacuum packed and the tightness of the cover will be augmented by the differential of atmospheric pressure pushing down the cover.

Crimping of the covers, as in the case of pop bottle caps which operate against positive internal pressure, also can make a gas-tight hermetic seal. But bottles fail more often than cans in becoming non-hermetic.

Protection of Food by Packaging Materials

Films and Foils; Plastics

Films and foils have different values for moisture and gas permeability, strength, elasticity, inflammability and resistance to insect penetration and many of these characteristics depend upon the film's thickness.

Important characteristics of the types of films and foils commonly used in food packaging are given in Table below. For the most part such films are used in the construction of inner containers. Since they are non-rigid, their main functions are to contain the product and protect it from contact with air or water vapour. Their capacity to protect against mechanical damage is limited, particularly when thin films are considered.

These materials can exist in many forms, depending upon such variables as identity and mixture of polymers, degree of polymerisation and molecular weight, spatial polymer orientation, use of plasticisers (softeners) and other chemicals, methods of forming such as casting, extrusion or calendering, etc.

One of the newer classes of plastic materials referred to as copolymers illustrate what can be done with mixtures of the basic units from which plastics are built. The term copolymer refers to a mixture of chemical species in the resin from which films and other forms can be made. The many variations possible make copolymers an important class of plastics to extend the range of useful food packaging applications.

Table: *Properties of packaging films*

Material	*Properties*
Paper	Strength; rigidity; opacity; printability.
Aluminium foil	Negligible permeability to water-vapour, gases and odours; grease proof, opacity and brilliant appearance; dimensional stability; dead folding characteristics.
Cellulose film (coated)	Strength; attractive appearance; low permeability to water vapour (depending on the type of coating used), gases, odours and greases; printability.
Polythene	Durability; heat-sealability; low permeability to water-vapour; good chemical resistance; good low-temperature performance.
Rubber hydrochloride	Heat-sealability; low permeability to water vapour, gases, odours and greases; chemical resistance.
Cellulose acetate	Strength; rigidity; glossy appearance; printability; dimensional stability.

Vinylidene chloride	Low permeability to water vapour, gases, copolymer odours and greases; chemical resistance; heat-sealability.
Polyvinyl chloride	Resistance to chemicals, oils and greases; heat-sealability.
Polyethylene terephtalate	Strength; durability; dimensional stability; low permeability to gases, odours and greases.

Source: FAD/WFP, (1970)

Plastic Sheets

- Cellophane paper can be used for packing of dried products, mainly for dried fruit leathers.
- Polyethylene sheets have a variety of uses. They are flexible, transparent and have a perfect resistance to low temperatures and impermeability to water vapour. An important advantage is that these sheets can be easily heat-sealed. Utilisation is in forms of sheets and bags. It is a good packing material for primary protection of dehydrated products. If a good protection is needed to prevent flavour and gas losses, it will be necessary to combine polyethylene with other materials.

Receptacles and Packagings in Plastic Materials

In this class there are three categories:

a. receptacles that can be heat treated: boxes, bottles and bags. Sterilisable bags used up to 120° C can be manufactured from same raw materials as described under plastic sheets and up to 100° C from cellophane. Polyethylene bags could be used to some extent for packing and pasteurization of sauerkraut.

b. receptacles that are not heat treated during processing of fruit and vegetables, also divided in bags and boxes. Bags are the most used type of packing from plastic materials and they are manufactured from polyethylene or cellophane; an important utilisation is for dried/ dehydrated fruits and vegetables.

c. special packagings - which can be contacted (Criovac type) by action of heat once the finished product is already inside the pack and the air is evacuated.

Laminates

Various flexible materials such as papers, plastic films, and thin metal foils have different properties with respect to water vapour transmission, oxygen permeability, light transmission, burst strength, pin holes and crease hole sensitivity, etc. and so multi-layers or laminates of these materials which combine the best features of each are used.

Commercial laminates containing up to as many as eight different layers are commonly custom designed for a particular product.

Laminations of different materials may be formed by various processes including bonding with a wet adhesive, dry bonding of layers with a thermoplastic adhesive, hot melt laminating where one or both layers exhibit thermoplastic properties, and special extrusion techniques. Such structured plastic films may be complete in themselves or be further bonded to papers or metal foils to produce more complex laminates.

Glass Containers

As far as food packaging is concerned, glass is chemically inert, although the usual problems of corrosion and reactivity of metal closures will of course apply. The principal limitation of glass is its susceptibility to breakage, which may be from internal pressure, impact, or thermal shock, all of which can be greatly minimised by proper matching of the container to its intended use and intelligent handling practices. Here consultation with the manufacturer cannot be over-stressed.

The heavier a jar or bottle for a given volume capacity the less likely it is to break from internal pressure. The heavier jar, however, is more susceptible to both thermal shock and impact breakage. Greater thermal shock breakage of the heavier jar is due to wider temperature differences which cause uneven stress between the outer and inner surfaces of the thicker glass. Greater impact breakage susceptibility of the heavier jar is due to the lower resilience of its thicker wall.

Coatings of various types can markedly reduce each of these types of breakage. These coatings, commonly of special waxes and silicones, lubricate the outside of the glass. As a result, impact breakage is lessened by bottles and jars glancing off one another rather than sustaining direct hits when they are in contact in high speed filling lines.

Surface coating after annealing protects glass surfaces from many of the minute scratches appearing in normal handling after annealing ovens; surface coating also improves the high gloss appearance of glass containers and is said to decrease the noise from glass to glass contact at filling lines.

With regard to thermal shock, it is good practice to minimise temperature differences between the inside and outside of glass containers whenever possible. Some manufacturers will recommend that a temperature difference of 44° C (80° F) between the inside and the outside not be exceeded. This requires slow warming of bottles before use for a hot fill and partial cooling before such containers are placed under refrigeration.

Classification

Glass used for receptacles in fruit and vegetable processing is a carefully controlled mixture of sand, soda ash, limestone and other materials made molten by heating to about 1500° C (2800° F).

Main classes of glass receptacles are:

a. jars which are resistant to heat treatments,

b. jars, glasses, etc. for products not submitted to heat treatment (marmalades, acidified vegetables, etc.);

c. glass bottles for pasteurized products (tomato juice, fruit juices, etc.) or not pasteurized (syrups) and

d. receptacles with higher capacity (flasks, etc.)

Jars for Sterilised/Pasteurized Canned Products

These receptacles may replace metal cans, taking into considering both the advantages and disadvantages they present. Advantages are: they do not react to food content; they are transparent and can be manufactured in various shapes; they use cheap raw materials and are reusable. Disadvantages: heavier than metal can of same capacity; fragile; lower thermal conductance and a limited resistance to thermal shocks.

Receptacles in this category must assure a perfect hermeticity after their pasteurization/sterilisation and cooling and this has to be achieved by the use of metallic (or glass) caps and specific materials for tightness. Taking into account the receptacles' closure method, there are two categories of receptacles:

a. glass jars with mechanical closure;

b. glass jars with pneumatic closure.

Jars for Products without Heat Treatment

For marmalades, jellies and jams glass jars with non hermetic closures using metal, glass or rigid plastic caps are used; however for these products the receptacles mentioned above may also be used.

The use of jars with pneumatic closure presents the advantage that some products (e.g. marmalades, jams) can be filled hot and therefore sterile in receptacles. Pneumatic closing generally protects against negative air action which is in this case eliminated from receptacles.

Glass Bottles

These receptacles are widely used both for

a) finished products which need pasteurization (ea. tomato juice, Knit juices, etc.) and for

b) those which are preserved as such (ea. fruit syrups).

Glass bottles in category a) are closed hermetically with metallic caps, provided with special materials for tightness. For glass bottles in category b) various corks, and aluminium caps with tightness materials may be used.

Glass Receptacles with High Capacity

In this category there are glass flasks with 3 and 10 litre capacity which can be hermetically closed by a SKO caps system and are resistant to product pasteurization (ea. tomato juice).

As bigger receptacles it is possible to use glass demijohns with usual capacity of 25 and 50 litre; these receptacles are used for preservation of fruit juices by warm process. Closing is performed with flexible rubber hoods.

Paper Packaging

As primary containers few paper products are not treated, coated or laminated to improve their protective properties. Paper from wood pulp and reprocessed waste paper will be bleached and coated or impregnated with such materials as waxes, resins, lacquers, plastics, and laminations of aluminium to improve water vapour and gas impermeability, flexibility, tear resistance, burst strength, wet strength, grease resistance, sealability, appearance, printability, etc.

Paper Sheets

- Kraft paper is the brown unbleached heavy duty paper commonly used for bags and for wrapping; it is seldom used as a primary container;
- parchment paper: acid treatment of paper pulp modifies the cellulose and gives water and oil resistance and considerable wet strength to this type of packaging material;
- glassine-type papers are characterised by long wood pulp fibres which impart increased physical strength;
- paper with plastic material sheets.

Receptacles from Paper or Cardboard

(paper = 8 to 150 g/m^2; cardboard = 150 to 450 g/m^2).

"Tin can"/tinplate

The "tin can" is a container made of tinplate.

Tinplate, a rigid and impervious material, consists of a thin sheet of low carbon steel coated on both sides with a very thin layer of tin. It can be produced by dipping sheets of mild steel in molten tin (hot-dipped tinplate) or by the electro-deposition of tin on the steel sheet (electrolytic tinplate). With the latter process it is possible to produce tinplate with a heavier coating of tin on one surface than the other (differentially coated).

Tin is not completely resistant to corrosion but its rate of reaction with many food materials is considerably slower than that of steel. The effectiveness of a tin coating depends on:

a. its thickness which may vary from about 0.5 to 2.0 µm (20 to 80 x 10(-6) in.);
b. the uniformity of this thickness;
c. the method of applying the tin which today primarily involves electrolytic plating;
d. the composition of the underlying steel base plate;
e. the type of food, and
f. other factors.

Some canned vegetables including tomato products actually owe their characteristics flavours to a small amount of dissolved tin, without which these products would have an unfamiliar taste. On the other hand, where tin reacts unfavourably with a particular food the tin itself may be lacquer coated.

The thickness of tinplate sheets may vary from 0.14 mm to 0.49 mm and is determined by weighing a sheet of known area and calculating the average thickness.

Tinplate sheets may be lacquered after fabrication to provide an internal or external coating to protect the metal surface from corrosion by the atmosphere or through reaction with the can contents.

They may also be printed by lithography to provide suitable instructions or information on containers fabricated from tinplate sheets (otherwise paper labels can be attached to the outer tinplate surface).

Under normal conditions the presence of the tin coating provides a considerable degree of electrochemical protection against corrosion, despite the fact that in both types of tinplate the tin coating is discontinuous and minute areas of steel base plate are exposed. With prolonged exposure to humid conditions, however, corrosion may become a serious problem.

The coatings not only protect the metal from corrosion by food constituents but also protect the foods from metal contamination, which can produce a host colour and flavour reactions depending upon the specific food. Particularly common are dark coloured sulphides of iron and tin produced in low acid foods that liberate sulphur compounds when heat processed, and bleaching of red plant pigments in contact with unprotected steel, tin, and aluminium.

***Table:** General types of can coatings*

Coating	*Typical uses*	*Type*
Fruit enamel	Dark coloured berries, cherries and other fruits requiring protection from metallic salts	Oleoresinous
C-enamel	Corn, peas and other sulphur-bearing products	Oleoresinous w. suspended zinc oxide
Citrus enamel	Citrus products and concentrates oleoresinous	Modified
Beverage can enamel	Vegetable juices; red fruit juices; highly corrosive fruits; non-carbonated beverages	Two-coated w. resinous base coat and vinyl top coat

Fruit Specific Preservation Technologies

Fruit Quality

Fruit quality goes back to tree stock, growing practices and weather conditions. Closer to the shipper and processor, however, are the degrees of maturity and ripeness when picked and the method of picking or harvesting.

There is a distinction between maturity and ripeness of a fruit. Maturity is the condition when the fruit is ready to eat or if picked will become ready to eat after further ripening. Ripeness is that optimum condition when colour, flavour and texture have developed to their peak.

Some fruit is picked when it are mature but not yet ripe. This is especially true of very soft fruit like cherries and peaches, which when fully ripe are so soft as to be damaged by the act of picking itself. Further, since many types of fruit continue to ripen off the tree, unless they were to be processed quickly, some would become overripe before they could be utilised if picked at peak ripeness.

From a technological point of view, fruit characterisation by species and varieties is performed on the basis of physical as well chemical properties: shape, size, texture, flavour, colour/pigmentation, dry matter content (soluble solids content), pectic substances, acidity, vitamins, etc. These properties are directly correlated with fruit utilisation. Table below shows which of the above mentioned properties have a major impact on the finished products obtained by fruit processing.

Table: *Optimal use of fruits as a function of their properties*

Processed	***Organoleptic (Sensory) Properties***				***Chemical Composition***		
Finished Products	***Shape***	***Texture***	***Flavour***	***Taste***	***Acidity***	***Sugars***	***Pectic Sust.***
Dried Fruits	+ +	+ +		+ +		+ +	
Fruit Juices			+ +	+ +	+ +		
Marmalade			+ +	+ +			+ +
Jams	+ +	+ +	+ +	+ +			
Jellies	+ +	+ +	+ +	+ +			+ +
Fruit Paste				+ +		+ +	

When to Pick

The proper time to pick fruit depends upon several factors; these include variety, location, weather, ease of removal from the tree (which change with time), and purpose to which the fruit will be put.

For example, oranges change with respect to both sugar and acid as they ripen on the tree; sugar increases and acid decreases. The ratio of sugar to acid determines the taste and acceptability of the fruit and the juice. For this reasons, in some countries there are laws that prohibit picking until a certain sugar-acid ratio has been reached.

In the case of much fruit to be canned, on the other hand, fruit is picked before it is fully ripe for eating since canning will further soften the fruit.

Quality Measurements

Many quality measurements can be made before a fruit crop is picked in order to determine if proper maturity or degree of ripeness has developed.

Colour may be measured with instruments or by comparing the colour of fruit on the tree with standard picture charts.

Texture may be measured by compression by hand or by simple type of plungers.

As fruit mature on the tree its concentration of juice solids, which are mostly sugars, changes. The concentration of soluble solids in the juice can be estimated with a refractometer or a hydrometer. The refractometer measures the ability of a solution to bend or refract a light beam which is proportional to the solution's concentration. A hydrometer is a weighted spindle with a graduated neck which floats in the juice at a height related to the juice density.

The acid content of fruit changes with maturity and affects flavour. Acid concentration can be measured by a simple chemical titration on the fruit juice.

But for many fruits the tartness and flavour are really affected by the ratio of sugar to acid. Percentage of soluble solids, which are largely sugars, is generally expressed in degrees Brix, which relates specific gravity of a solution to an equivalent concentration of pure sucrose.

In describing the taste of tartness of several fruits and fruit juices, the term "sugar to acid ratio" or "Brix to acid ratio" are commonly used. The higher the Brix the greater the sugar concentration in the juice; the higher the "Brix to acid ratio" the sweeter and lees tart is the juice.

Harvesting and Preprocessing

The above and other measurements, plus experience, indicate when fruit is ready for harvesting and subsequent processing.

A large amount of the harvesting of most fruit crops is still done by hand; this labour may represent about half of the cost of growing the fruit. Therefore, mechanical harvesting is currently one of the most active fields of research for the agricultural engineer, but also requires geneticists to breed fruit of nearly equal size, that matures uniformly and that is resistant to mechanical damage.

A correct manual harvesting includes some simple but essential rules:

- the fruit should be picked by hand and placed carefully in the harvesting basket; all future handling has to be performed carefully in order to avoid any mechanical damage;
- the harvesting basket and the hands of the harvester should be clean;
- the fruit should be picked when it is ready to be able to be processed into a quality product depending on the treatment which it will undergo.

It is worth emphasising the fact that the proximity of the processing centre to the source of supply for fresh raw materials presents major advantages; some are as follows:

- possibility to pick at the best suitable moment;
- reduction of losses by handling/transportation;
- minimises raw material transport costs;
- possibility to use simpler/cheaper receptacles for raw material transport.

Once it has left the tree, the organoleptic properties, nutritional value, safety and aesthetic appeal of the fruit deteriorates in varying degrees. The major causes of deterioration include the following:

a. growth and activity of micro-organisms;

b. activities of the natural food enzymes;

c. insects, parasites and rodents;

d. temperature, both heat and cold;
e. moisture and dryness;
f. air and in particular oxygen;
g. light and
h. time.

The rapidity with which foods spoil if proper measures are not taken is indicated in Table below.

***Table:** Useful storage life of some food products*

Food Products	*Generalized Storage Life (Days) at 21°C (70°F)*
Animal Flesh, Fish, Poultry	1-2
Dried, salted, smoked meat and fish	360 and more
Fruits	1-7
Dried Fruits	360 and more
Leafy Vegetables	1-2
Root Crops	7-20

Source: Desrosier and Desrosier (1977)

Reception - Quality and Quantity

Fruit reception at the processing centre is performed mainly for following purposes:

- checking of sanitary and freshness status;
- control of varieties and fruit wholeness;
- evaluation maturity degree;
- collection of data about quantities received in connection to the source of supply: outside growers/farmers, own farm.

Variety control is needed in order to identify that the fruit belongs to an accepted variety as not all are suitable for different technological processes.

Fruit maturity degree is significant as industrial maturity is required for some processing/preservation methods while for others there is the need for an edible maturity when the fruit has full taste and flavour.

Special attention is given to size, appearance and uniformity of fruit to be processed, mainly in the form of fruit preserved with sugar using whole/half fruits ("with fruit pieces").

Some laboratory control is also needed, even if it not easy to precisely establish the technological qualities of fruit because of the absence of enough reliable rapid analytical methods able to show eventual deterioration. The only reliable method for evaluating the quality is the combination of data obtained through organoleptic/taste controls and by simple analytical checks which are possible to perform in a small laboratory: percentage of soluble solids by refractometer, consistency/texture measured with simple penetrometers, etc.

Temporary Storage before Processing

This step has to be as short as possible in order to avoid flavour losses, texture modification, weight losses and other deterioration that can take place over this period.

Some basic rules for this step are as follows:

- keep products in the shade, without any possible direct contact with sunlight;
- avoid dust as much as possible;
- avoid excessive heat;
- avoid any possible contamination;
- store in a place protected from possible attack by rodents, insects, etc.

Cold storage is always highly preferred to ambient temperature. For this reason a very good manufacturing practice is to use a cool room for each processing centre; this is very useful for small and medium processing units as well.

Washing

Harvested fruit is washed to remove soil, micro-organisms and pesticide residues.

Fruit washing is a mandatory processing step; it would be wise to eliminate spoiled fruit before washing in order to avoid the pollution of washing tools and/or equipment and the contamination of fruit during washing.

Washing efficiency can me gauged by the total number of micro-organisms present on fruit surface before and after washing - best result are when there is a six fold reduction. The water from the final wash should be free from moulds and yeast; a small quantity of bacteria is acceptable.

Fruit washing can be carried out by immersion, by spray/showers or by combination of these two processes which is generally the best solution: pre-washing and washing.

Some usual practices in fruit washing are:

- addition of detergents or 1.5% HCl solution in washing water to remove traces of insect-fungicides;
- use of warm water (about 50°C) in the pre-washing phase;
- higher water pressure in spray/shower washers.

Washing must be done before the fruit is cut in order to avoid losing high nutritive value soluble substances (vitamins, minerals, sugars, etc.).

Sorting

Fruit sorting covers two main separate processing operations:

a. removal of damaged fruit and any foreign bodies (which might have been left behind after washing);

b. qualitative sorting based on organoleptic criteria and maturity stage.

Mechanical sorting for size is usually not done at the preliminary stage. The most important initial sorting is for variety and maturity.

However, for some fruit and in special processing technologies it is advisable to proceed to a manual dimensional sorting (grading).

Trimming and Peeling (Skin Removal)

This processing step aims at removing the parts of the fruit which are either not edible or difficult to digest especially the skin.

Up to now the industrial peeling of fruit and vegetables was performed by three procedures:

a. mechanically;

b. by using water steam;

c. chemically; this method consists in treating fruit and vegetables by dipping them in a caustic soda solution at a temperature of 90 to 100° C; the concentration of this solution as well as the dipping or immersion time varying according to each specific case.

Cutting

This step is performed according to the specific requirements of the fruit processing technology.

Heat Blanching

Fruit is not usually heat blanched because of the damage from the heat and the associated sogginess and juice loss after thawing. Instead, chemicals are commonly used without heat to inactivate the oxidative enzymes or to act as antioxidants and they are combined with other treatments.

Ascorbic/Citric Acid Dip

Ascorbic acid or vitamin C minimises fruit oxidation primarily by acting as an antioxidant and itself becoming oxidised in preference to catechol-tannin compounds. Ascorbic acid is frequently used by being dissolved in water, sugar syrup or in citric acid solutions.

It has been found that increased acidity also helps retard oxidative colour changes and so ascorbic acid plus citric acid may be used together. Citric acid further reacts with (chelates) metal ions thus removing these catalysts of oxidation from the system.

Sulphur Dioxide Treatment

Sulphur dioxide may function in several ways:

- sulphur dioxide is an enzyme poison against common oxidising enzymes;
- it also has antioxidant properties; i.e., it is an oxygen acceptor (as is ascorbic acid);
- further SO2 minimises non enzymatic Maillard type browning by reacting with aldehyde groups of sugars so that they are no longer free to combine with amino acids;
- sulphur dioxide also interferes with microbial growth.

In many fruit processing pre-treatments two factors must be considered:

a. sulphur dioxide must be given time to penetrate the fruit tissues;

b. SO_2 must not be used in excess because it has a characteristic unpleasant taste and odour, and international food laws limit the SO_2 content of fruit products, especially of those which are consumer oriented (e.g. except semi-processed products oriented to further industrial utilisation).

Commonly a 0.25 % solution (except for semi-processed fruit products which are industry oriented and use a 6% solution) of SO_2 or its SO_2 equivalent in the form of solutions of sodium sulphite, sodium bisulphite or sodium/ potassium metabisulphite are used.

Fruit slices are dipped in the solution for about two to three minutes and then removed so as not to absorb too much SO2. Then the slices are allowed to stand for about one to two hours so that the SO2 may penetrate throughout the tissues before processing.

Sulphur dioxide is also used in fruit juice production to minimise oxidative changes where relatively low heat treatment is employed so as not to damage delicate juice flavour. Dry sulphuring is the technological step where fruit is exposed to fumes of SO2 from burning sulphur or from compressed gas cylinders; this treatment could be used in the preparation of fruits (and some vegetables) prior to drying/dehydration.

Sugar Syrup

Sugar syrup addition is one of the oldest methods of minimising oxidation. It was used long before the causative reactions were understood and remains today a common practice for this purpose.

Sugar syrup minimises oxidation by coating the fruit and thereby preventing contact with atmospheric oxygen.

Sugar syrup also offers some protection against loss of volatile fruit esters and it contributes sweet taste to otherwise tart fruits.

It is common today to dissolve ascorbic acid and citric acid in the sugar syrup for added effect or to include sugar syrup after an SO_2 treatment.

Fresh Fruit Storage

Some fruit species and specially apples and pears can be stored in fresh state during cold season in some countries' climatic conditions.

Fruit for fresh storage have to be autumn or winter varieties and be harvested before they are fully mature. This fruit also has to be sound and without any bruising; control and sorting by quality are mandatory operations.

Sorting has to be carried out according to size and weight and also by appearance; fruit which is not up to standard for storage will be used for semi-processed product manufacturing which will be submitted further to industrial processing.

Harvested fruit has to be transported as soon as possible to storage areas. Leaving fruit in bulk in order to generate transpiration is a bad practice as this reduces storage time and accelerates maturation processes during storage.

In order to store large quantities of fruit, silos have to be built.

Technology of Semi-processed Fruit Products

The semi-processed fruit products are manufactured in order to be delivered to industry processing centres (in the fruit producing country itself or in importing countries) where they will be further manufactured in consumer oriented finished products: jams, jellies, syrups, fruits in syrup, etc.

In the practice of semi-processed fruit products and for the purpose of this document the following categories are defined:

a. fruit "pulps": semi-processed products, not refined, obtained by mechanical treatment (or, less often, by thermal treatment) of fruit followed by their preservation. Either whole fruit, halves or big pieces are used which enables easy identification of the species. "Pulps" can be classified in boiled or non boiled (raw).

b. fruit "purées-marks": semi-processed products obtained by thermal and mechanical treatment or, very rare, raw and then refined, operations by which all nonedible parts (cores, peels, etc.) are removed. "Purées-marks" are classified in boiled (the more usual case) and non boiled (raw).

c. semi-processed juices: products obtained by cold pressure or very rare by other treatments (diffusion, extraction, etc.) followed by the preservation.

Technical Processes for Preservation of Semi-processed Fruit Products

Preservation can be achieved by chemical means, by freezing or by pasteurization. The choice of preservation process for each individual case is a function of the semi-processed product type and the shelf life needed.

Chemical preservation. In many countries, in practice, this is carried out with sulphur dioxide, sodium benzoate, formic acid and, on a small scale, with sorbic acid and sorbates.

Preservation with sulphur dioxide is a widespread process because of its advantages: universal antiseptic action and very economic application. The drawbacks of SO_2 are: SO_2 turn firms the texture of some fruit species (pomaces), desulphiting is not always complete and recolouring of red fruits is not always complete after desulphitation.

Practical preservation dosage levels with SO_2 for about 12 months is 0.18-0.20% SO_2 (with respect to the product to be preserved).

This level could be reduced to 0.09% SO_2 for 3 months and to 0.12% SO_2 for 6 months preservation.

The preservation with sulphur dioxide is in use mainly for "pulps" and for "purées-marks".

Chemical preservation can be performed from a practical point of view by the utilisation of 6% SO2 water solutions or by direct introduction of sulphur dioxide gas in the product (for "purées-marks"). The preparation of 6% SO2 solutions is done by bubbling the gas from cylinders in cold water; from a 50 kg SO_2 compressed gas cylinder results 830 l of 6% SO2 solution.

These SO_2 solutions have to be stored in cool places, in closed receptacles and with periodic concentration control/check by titration or by density measurements approximate results.

Preservation with sodium benzoate has the following advantages: it does not firm up the texture and does not modify fruit colour. The disadvantages are: it is not a universal antiseptic, its action needs an acid medium and the removal is partial. Sodium benzoate is in use for "pulps" and for "purées-marks" but less for fruit juices.

Practical dosage level for 12 months preservation is 0.18-0.20 % sodium benzoate, depending on the product to be preserved. Sodium benzoate is used as a solution in warm water; the dissolution water level has to be at maximum 10% reported to semi-processed product weight.

Formic acid preservation is performed mainly for semi-processed fruit juices at a dosage level of 0.2 % pure formic acid (100%). Formic acid is an antiseptic effective against yeasts, does not influence colour of products and is easily removed by boiling.

Formic acid could be diluted with water in order to insure a homogeneous distribution in the product to be preserved; water has to be at maximum 5 % of the product weight. Because of a potential effect of pectic substance degradation, formic acid is less in use for "pulps" and "purées-marks" preservation.

Sorbic acid used as potassium sorbate (easily water soluble) can be used for preservation of fruit semi-processed products at a dosage level of 0. 1% maximum. Advantages of sorbates are: they are completely harmless and without any influence on the organoleptic properties of semi-processed fruit products.

Preservation by pasteurization. As fruit has a low pH, preservation of semiprocessed fruit products could also be performed by pasteurization (heat treatment step at maximum temperature of 100° C), the length of this step varying with the size of the receptacles.

The advantages of this type of treatment are: hygienic process, which assure a long term preservation; the disadvantages are: need for air tight receptacles, and pectic substances could begin to deteriorate if the thermal treatment is too long.

Thermal preservation of fruit semi-processed products could also be done by a "selfpasteurization": very hot semi-processed products are filled into receptacles (e.g. metal cans) which are sealed and then inverted in order to sterilise the air which goes through the hot fruit mass.

Preservation by freezing. This is done on an industrial scale in some countries and can be done with or without sugar addition. The advantages of this process are: absence of added substances; very good preservation of quality of fruit constituents (pectic substances, vitamins, etc.) and good preservation of organoleptic properties (flavour, taste, colour). Freezing is done at about -20 to -30° C and storage at -10 to -18° C.

Freezing is applied mainly to semi-processed fruit products aimed at very high quality and high cost finished products.

Technological flow-sheet for semi-processed fruit "pulps": chemical preservation.

Sorting is needed in order to remove sub-standard fruit (with moulds, with diseases, etc.) and all foreign bodies.

Washing is obligatory in order to remove all impurities which cannot be eliminated at the processing step in finished products.

Coring and Cutting, mainly for pomace fruits, has as main objective a better utilisation of preservation "space" in receptacles and is not mandatory; this will be defined by customer/supplier agreements/standards. This operation is preferably performed by mechanical means.

Preservation is carried out with the 6% SO2 solution which is added to the prepared fruits (placed in bulk in receptacles) in the quantity needed to obtain the preservation dosage level. For a better/homogeneous preservative distribution, the initial 6% SO2 solution could be diluted with water; however, the diluted solution (which will be filled in receptacles) has to be at a dosage level of less than 10% of the semi-processed product weight.

For some soft fruit, especially strawberries, preservation is done with a mix of 6% SO2 solution and calcium bisulphite solution (containing also 6% SO2).

Preparation of calcium bisulphite solution is done by the introduction of 30 kg of CaO in 1 m^3 SO2 solution and mixing up to clarification. The resulting solution is mixed with the initial 6% SO2 solution, generally in a 1:1 ratio, but the ratio can be adapted to the fresh fruit texture. Firming of soft fruit texture by this treatment is based on the formation of calcium pectate with pectic substances from fruit tissues.

In the case of sodium benzoate, formic acid or potassium sorbate, the dosage levels to be used are as indicated above with the rule that it is not allowed to add more that 10% liquid in receptacles on the prepared fruits.

Preservation by pasteurization or "self-pasteurization" will need as additional steps: a) boiling with a minimum water addition (maximum 10%); b) filling of receptacles; c) hermetic closing followed by d) pasteurization or "self-pasteurization".

Some general technical data for the preparation of chemical preserved semi-processed fruit "pulps" are seen in Table below.

Table: *General technical processing data for semi-processed fruit "pulps"*

Fruit species	Preliminary operations	Preservation means
Apples, pears, quinces	Sorting, washing, coring, cutting	Sulphur dioxide
Prunes, peaches, wax cherries, apricots	Sorting, washing, stone removal (pitting)	Sulphur dioxide or sodium benzoate
Cherries	Sorting, washing	Sulphur dioxide or sodium benzoate, sometimes with calcium bisulphite addition
Strawberries	Sorting, washing	Sulphur dioxide in mix with calcium bisulphite
Wild berries	Sorting, washing	Sulphur dioxide or sodium benzoate; in some cases with calcium bisulphite addition

Technological Flow-sheet for Semi-processed "Purée-marks"

The general technological flow-sheet includes the following operations:

Sorting and Washing are obligatory and are carried out in a similar manner as for "pulps".

Heat Treatment/Boiling is needed in order to soften the fruit tissues before refining. For some fruits as strawberry and wild berries, this step is not done and fruits are refined "raw" in order to preserve their flavour.

Pulping is performed with specific equipment - refiners or pulpers - which eliminate seeds, pits and other non edible parts (peels, cores, etc.).

Preservation is carried out by chemical means, by freezing or by pasteurization.

The general technical/processing data for the manufacture of "purée-marks" are seen in Table below.

Table: *General technical processing data for semi-processed "purées - marks"*

Fruit species	Preliminary operations	Preservation means
Apples, pears, quinces	Sorting, washing, boiling, refining	Sulphur dioxide; in less frequent cases formic acid or sodium benzoate
Prunes, peaches, wax cherries, apricots,	Sorting, washing, boiling, refining	Sulphur dioxide, formic acid or sodium benzoate; cherries freezing with or without sugar; self-pasteurization
Strawberry, wild berries	Sorting, washing, refining	Chemical preservation, freezing or self pasteurization

Fruit Sugar Preserves Technology; Jams, Jellies, Marmalade, Fruit Paste

As a overall rule of thumb, a sugar concentration of about 60% in finished or processed fruit products generally insures their preservation. Preservation is not only determined by the osmotic pressure of sugar solutions but also by the water activity values in the liquid phase, which can be lowered by sugar addition; and by evaporation down to 0.848 aw; this value however does not protect products from mould and osmophile yeast attack.

Maximum saccharose concentration that can be achieved in the liquid phase of the product is 67.89%; however higher total sugar quantities (up to 70-72%) found in products are explained by an increased reducing sugar solubility resulting from saccharose inversion.

Jams

The preservation of fruit by jam making is a familiar process carried out on a small scale by housewives in many parts of the world. Factory jam making has become a highly complex operation, where strict quality control procedures are employed to ensure a uniform product, but the manufacturing operations employed are in essence the same as those employed in the house.

Fresh or pre-cooked fruit is boiled with a solution of cane or beet sugar until sufficient water has been evaporated to give a mixture which will set to a gel on cooling and which contains 32-34% water.

Gel formation is dependent on the presence in the fruit of the carbohydrate pectin, which at a pH of 3.2 - 3.4 and in the presence of a high concentration of sugar, has the property of forming a viscous semi-solid.

During jam boiling, all micro-organisms are destroyed within the product, and if it is filled hot into clean receptacles which are subsequently sealed, and then inverted so that the hot jam contacts the lid surface, spoilage by micro-organisms will not take place during storage.

The composition of jam made from stone fruit and berry fruit is shown in Table below. About 30% of the vitamin C present in fresh fruit is destroyed during the jam-making process, but that which remains in the finished product is stable during storage.

The high moisture content of jam (equivalent to an equilibrium relative humidity of about 82%) makes it susceptible to mould damage once the receptacle has been opened and exposed from some time to the air. No problems of microbiological spoilage are likely to arise in the canned product during storage.

***Table:** Composition of some fruit jams*

Type of Jam	Moisture content %	Sugar (as invert sugar, %)	Vitamin C mg/100 g
Jam made from berry fruits: strawberry, raspberry, etc.	29.8	69.0	10 - 25
Jam made from stone fruits: apricot, peach, etc.	29.6	69.3	10 - 35

Source: FAD/WFP, 1970

Marmalade

This sugar preserve is defined as "semisolid or gel-like product prepared from fruit ingredients together with one or more sweetening ingredients and may contains suitable food acids and food pectins; the ingredients are concentrated by cooking to such a point that the TSS - Total Soluble Solids - of the finished marmalade is not below 65%".

Fruit Paste

Fruit paste is a product obtained in the same way as special non-gelified fruit marmalade but with a lower water content - about 25% TSS in fruit paste.

Lowering water content could be achieved by continuing boiling of the product or by drying the product by natural or artificial drying. An example of paste without sugar is the sun dried apricot or prune paste.

General procedure for the preparation of jams, jellies and marmalade:

a. Boil the pulp or the juice (with water when necessary)
b. Add the pectin
   - to the batch while stirring very vigorously
   - Pectin which has previously been mixed with 5 times its weight in sugar taken from the recipe)
c. Boil for about 2 minutes to assure a complete dissolution
d. Add the sugar while keeping the batch boiling
e. Boil down quickly to desired Brix
f. Add the acid (usually citric acid) and remove the froth
g. Fill hot into the (previously cleaned) jars and close
h. Invert the jars for three minutes to pasteurize the cover.

• Note: the pectin in solution can also be added at the end of the step (e) and has to be prepared as follows: use a strong blender. For one litre of water add slowly into the blender 25 g of pectin mixed with 100 g of sugar taken from the recipe.

Basic recipes

The following recipes must be considered only as guidelines because the composition of the fruit can vary; also the taste of the consumers varies concerning the consistency, the sweetness and acidity.

Before starting to make jam it is important to know the yield to settle the question on containers. The calculation is made as follows:

$$\frac{\text{Total Soluble Solids content of all the raw ingredients}}{\text{Percentage of Total Soluble Solids in the final product}} \times 100$$

In these basic recipes it is assumed that the fruits are poor in pectin content.

Recipe 1. Fruit: sugar = 50:50; desired Brix in the finished product is 68.

Soluble Solids:

10 kg of fruit at 10% TSS	1.000 kg
10 kg of sugar	10.000 kg
60 g of pectin (grade 200)	0.060 kg
55 g of citric acid	0.055 kg
	11.115 kg

$$\text{Yield} = \frac{11.115 \times 100}{68} = 16.4 \text{ Kg}$$

Recipe 2. Fruit: sugar = 45:55; desired Brix in the finished product is 68.

Soluble Solids

d10 kg of fruit at 10% TSS	1.000 kg
2.5 litre of water	-
12.2 kg of sugar	12.200 kg
65 g of pectin grade 200	0.060 kg
60 g of citric acid	0.060 kg
	13.325 kg

$$\text{Yield} = \frac{13.325 \times 100}{68} = 19.6 \text{ Kg}$$

Recipe 3. Fruit: sugar = 40:60; desired Brix in the finished product is 68.

Soluble Solids

d10 kg of fruit at 10% TSS	1.000 kg
3.3 litre of water	-
15 kg of sugar	15.000 kg
85 g of pectin grade 200	0.085 kg
80 g of citric acid	0.080 kg
	16.165 kg

$$\text{Yield} = \frac{16.165 \times 100}{68} = 23.8 \text{ Kg}$$

Various factors must be taken into account:

1. Size of the container: the quantity of pectin indicated in the recipes is valid for containers of 1 kg or less.

If container capacity is between:	*Increase pectin by*
1 kg and 2.5 kg	5%
2.5 kg and 5.0 kg	10%
5.0 kg and 10.0 kg	20%
10.0 kg and 20.0 kg	30%

2. Finishing point: the quantity of pectin is given for a final Brix - Total Soluble Solids (TSS) of 68%.

   If the final Brix is 66 increase the pectin by 5%

   " " is 65 " " " by 10%

   is 64 " " " by 15%

   " " is 62 " " " by 20%

   " " is 60 " " " by 30%

3. Acidic taste. If the product is too acid, replace the citric acid by tartaric acid (63% of the amount of citric acid).
4. Formation of clots: If batch clots, it is probably due to the pH being too low or the or TSS being too high; correct accordingly.
5. Formation of liquid at the surface: if liquid forms on the surface, it is probably due to too low a pH or too low pectin content.
6. Crystallisation: a) if liquid forms on the surfaces, then the pH is too low; reduce the acid content; b) if liquid does not form on the surface, then TSS or pH is too high.
7. Formation of mould: the TSS is probably below 68 deg. Brix. The filling may have been done at a low temperature. If the containers are large, wait until they are cold before closing.
8. Wrong batch: dilute the jam with water to 30% TSS; cook briefly. Add this diluted jam to a new batch but in a ratio not exceeding 10%.

Processing of Pineapple-papaya Jam

The fruit should be prepared as per previous instructions. For pineapples, the ends are removed and discarded; the cores and outer parts of the fruits are also removed. The fruit cylinders obtained are pulped through a special extractor (Fitzpatrick communiting machine) equipped with a 0.40-in screen sieve; the pulp thus obtained is used for making jam.

The papaya are prepared by hand-peeling the fruit; the fruit is then halved and the seeds removed. It is then pulped in the communiting machine using a 0.40-in screen sieve. When ginger root is used as flavouring, it is peeled and macerated in a Kenwood blender to a very fine consistency.

A typical formula for a pineapple-papaya jam (50:50 ratio) with ginger flavouring is given as follows:

Pineapple pulp	25.0
Papaya pulp	25.0 pounds
Cane sugar	50.0
Apple pectin (150 grade)	6.0 ounces
Citric acid	6.4
Fresh ground ginger	7.5

Processing is carried out in the following way:

The weighed fruit pulp is placed in a stainless steel steam-jacketed kettle and heated to about 110°F under constant stirring.

When the product reaches this temperature, the heat is turned off. The pectin (mixed in about ten times its weight with some of the weighed sugar), is then mixed into the fruit pulp, stirring constantly in order to prevent the pectin from clotting.

When the pectin has dissolved, the remainder of the sugar is added and dissolved completely in the mixture. The heat is then turned on and the jam mixture is stirred constantly until it starts boiling vigorously. During the remainder of the cooking, the product is stirred occasionally. Near the finishing point (approximately 221° F), the citric acid and the ginger (if it is used) are also added.

Determination of the finishing point is done by removing samples at intervals, cooling, and reading the soluble solids by means of a refractometer equipped with a Brix scale. After the jam reaches the proper Total Soluble Solids content, the heat is turned off and the surface scum/foam is removed.

The jam then is quickly put into receptacles which have been cleaned and sterilised with boiling in water for 30 minutes. The filling operation is done rapidly in order to prevent the temperature of the jam from falling below 190° F.

After filling, sterilised lids (boiled for 30 minutes in water) are placed on the receptacles and they are then sealed.

After this operation the receptacles are inverted for about 3 minutes to insure that the lids are sterilised. The receptacles are then placed upright.

At this stage it is not necessary to do any further processing, therefore the receptacles are cooled in running cold water until they reach a temperature slightly above room temperature. They are then dried in air and labelled.

Evaluation of Finished Products

During production at medium/large scale, it is recommended that quality controls be performed during manufacturing. After ten weeks of storage at room temperature it is recommended that an examination of finished products be performed. The receptacles are opened and contents carefully emptied on to enamel trays without disturbing the formation of the jam.

The empty cans (if metal cans were used) are then inspected for signs of corrosion. Factors other than flavour include colour, appearance, syrup separation, firmness and spreading quality. For flavour, jam is tested on pieces of bread. Samples are taken for measurement of pH (with a glass electrode pH meter) and Total Soluble Solids (with a refractometer equipped with a Brix scale).

This evaluation enables to have a quality check during product shelf life and to obtain data needed for necessary improvements of future productions.

For pineapple-papaya jam, products made with 30% pineapple and 70% papaya with added ginger has the highest score for flavour. The use of plain tin cans causes corrosion problems which is not the case when acid resistant lacquer cans are used.

Pineapple Jam Making

1. Boil 40 lb. of pulp and 12 lb. of water.
2. Add 225 g of pectin to the batch while stirring rapidly.
3. Boil for about 90 sec to assure complete dissolution.
4. Add 60 lb. of sugar gradually if possible in several portions, while keeping the batch boiling.
5. Boil down quickly to 69 deg. Brix (223 deg. F).
6. Take off steam, remove foam.
7. Add 300 cc citric acid solution 50%.
8. Fill hot (180 deg. F).
9. Invert receptacles for 3 minutes.

Check each batch for Brix 68-70 deg.; acidity/pH = 3.2 +/- 0.2.

Evaluate: absence of defects; colour; flavour; consistency.

Various fruit: sugar ratios can be manufactured; some basic recipes are as follows:

	Ratio 50:50	Ratio 45:55	Ratio 40:60
Fruit	55 lb	49.5 lb	44 lb
Water	-	11 lb	13.2 lb
Sugar	55 lb	60.5 lb	66 lb
Pectin (150 grade)	225 g	237.5 g	250 g
Citric acid (50% sol.)	300 cc	320 cc	335 cc

Fruit Juice Technologies

Fruit juices are products for direct consumption and are obtained by the extraction of cellular juice from fruit, this operation can be done by pressing or by diffusion.

For the purpose of this document, the technology of fruit juice processing will cover two finished product categories:

- juices without pulp ("clarified" or "not clarified");
- juices with pulp ("nectars").

We will also define as

a. "natural juices" products obtained from one fruit; and
b. "mixed juices" products obtained from the mix of two or three juices from different fruit species or by adding sugar.

Juices obtained by removal of a major part of their water content by vacuum evaporation or fractional freezing will be defined as "concentrated juices".

Technological Steps for Processing of Fruit Juices Without Pulp

Fruit juices must be prepared from sound, mature fruit only.

Soft fruit varieties such as grapes, tomatoes and peaches should only be transported in clean boxes which are free from mould and bits of rotten fruit.

Washing: fruit must be thoroughly washed. Generally, fruit will be submitted to a pre-washing before sorting and a washing step just after sorting.

Sorting: removal of partially or completely decayed fruit is the most important operation in the preparation of fruit for production of first quality fruit juices; sorting is carried out on moving inspection belts or sorting tables.

Crushing/Grinding/Disintegration Step is applied in different ways and depends on fruit types:

Crushing for grapes and berries;

Grinding for apples, pears;

Disintegration for tomatoes, peaches, mangoes, apricots.

This processing step will need specific equipment which differs from one type of operation to another.

Enzyme Treatment of crushed fruit mass is applied to some fruits by adding 2-8% pectolitic enzymes at about 50° C for 30 minutes.

This optional step has the following advantages: extraction yield will be improved, the juice colour is better fixed and finished product taste is improved.

However, for fruit which is naturally rich in pectic substances, this treatment makes the resulting "exhausted" material useless for industrial pectin production.

Heating of crushed fruit mass before juice extraction is an optional step used for some fruit in order to facilitate pressing and colour fixing; at same time, protein coagulation takes place.

Pressing to extract juice.

Diffusion is an alternative step for juice extraction and can be carried out discontinuously or in batteries at water temperature of about 80-85 ° C.

Juice Clarifying can be performed by centrifugation or by enzyme treatment. Centrifugation achieves a separation of particles in suspension in the juice and can be considered as a pre-clarifying step. This operation is carried out in centrifugal separators with a speed of 6000 to 6500 RPM.

Enzyme clarifying is based on pectic substance hydrolysis; this will decrease the juices' viscosity and facilitate their filtration. The treatment is the addition of pectolitic enzyme preparations in a quantity of 0.5 to 2 g/l and will last 2 to 6 hours at room temperature, or less than 2 hours at 50° C, a temperature that must not be exceeded.

The control of this operation is done by checking the decrease in juice viscosity. Sometimes, the enzyme clarifying is completed with the step called "sticking" by the addition of 5-8 g/hl of food grade gelatine which generates a flocculation of particles in suspension by the action of tannins.

Filtration of clarified juice can be carried out with kieselgur and bentonite as filtration additive in press-filters (equipment).

De-tartarisation is applied only to raisin juice and is aimed to eliminate potassium bitartrate from solution. This step can be performed by the addition of 1% calcium lactate or 0.4% calcium carbonate.

Pasteurization of juice can be done for temporary preservation (pre-pasteurization) and in this case this operation is carried out with continuous equipment (heat exchangers, etc.); warm juice is stored in drums or large size receptacles (20-30 kg). Pasteurization conditions are at 75°C in continuous stream.

Pasteurization of bottled juice is then carried out just before delivery to the market; this is performed in water baths at 75° C until the point where the juice reaches 68° C. In cases when the final pasteurization is done without pre-pasteurization and temporary storage, modern methods use a rapid pasteurization followed by aseptic filling in receptacles.

Rapid pasteurization conditions are as follows: temperature about 80° C, over 10-60 sec., followed by cooling; all operations are carried out in continuous stream.

Preservation under CO2 pressure may be done at a concentration of 1.5% CO2 under a pressure of 7 kg/cm^2. At the distribution step, proceed at CO2 decompression and the juice is then submitted to a sterilising filtration and aseptic filling in receptacles.

Preservation by freezing is carried out at about -30° C, after a preliminary de-aeration; storage is at -15 to -20° C.

Production of concentrated juices by evaporation is performed under vacuum (less than 100 mm Hg residual pressure) up to a concentration of 65-70% total sugar which assures preservation without further pasteurization. Modern evaporation installations recover flavours from juices which are then reincorporated in concentrated juices.

Additional operations for juice manufacturing are the vacuum de-aeration and mixing with other fruit juices or with sugar.

For the production of non clarified juices the centrifugation is the only specific step, enzyme clarifying and subsequent filtration being eliminated.

The optimum sugar/acid ratio for the majority of fruit, mainly for pomaces, is 10/1 to 15/1.

Fruit which is rich in carotenoids (apricots, peaches, etc.) is only processed as juices with pulp ("nectars").

Technological Flow-sheet for Fruit Juices with Pulp ("Nectars")

This process is divided at industrial scale in two categories of operations:

a. processing for obtaining juices;

b. juice conditioning for preservation.

Operations in the first category are differ according to the type of fruit which to be processed. Pomaces (apples, pears) are washed and sorted and then crushed in a colloid mill; fruit purée is then passed through a screw type heating equipment where direct steam is used as a source of heat. Warm fruit mass is treated in a pulper with a 2 mm screen and then through an extractor similar with the equipment used for tomato juice.

Stone fruits (apricots, peaches, cherries, etc.) after washing and sorting are submitted to steam in a continuous heater, then the warm fruit mass is passed through a pulper and then an extractor (as mentioned above). Berries (strawberry, wild berries, etc.) are washed, sorted and then crushed, preheated and then introduced in extractor.

In order to avoid browning and undesirable taste modifications it is usual to add about 0.05% ascorbic acid.

Second category type of operations are similar for all fruit species. Partial elimination of cellulose is achieved with a continuous centrifugal separator; the resulting juice is then processed in order to adjust sugar and acid content for viscosity.

Sugar (about 8-10%) is added as a syrup (in water or in the juice of same fruit obtained by pressure). Acidity is adjusted with citric or tartaric acid. The adjusted juice is then deaerated under vacuum at about 40° C. This step aims at avoiding oxidative reactions and vitamin C loss reduction.

An important subsequent step is an intensive homogenisation (under pressure at 150-180 A) in order to obtain particles with dimensions below 100. The homogenised juice obtained is then continuously pasteurized in plate heat exchanger equipment at a temperature of about 130° C, cooled down to about 90° C and aseptically packed in receptacles.

The principal characteristics of fruit "nectars" are uniformity and stability of the content provided by the advanced disintegration of fruits. Stability can be obtained by increasing product viscosity by adding pectin for fruit which is deficient in this component. In order to avoid "separation", intensive homogenisation is carried out as described above.

Fruit "nectars" contain all the important components of the original fruit and to a large extent maintain their taste and flavour. The sugar/acidity (as citric acid) ratio is to a large extent determined by the type of fruit and the correction applied; for example, this ratio is 30 for apricots, 40 for peaches, 160 for pears, etc.

Banana and Plantain Processing Technologies

Products: Uses and Dietary Significance

Most of the world's bananas are eaten either raw, in the ripe state, or as a cooked vegetable, and only a very small proportion are processed in order to obtain a storable product. This is true both at a traditional village level with both dessert and cooking bananas and when considering the international trade in dessert bananas. In general, preserved products do not contribute significantly to the diet; however, in some localised areas the products are important in periods when food are scarce. Probably the most widespread and important product is flour preparation from unripe banana and plantains by sun-drying. In Uganda, dried slices known as "mutere" are prepared for storage from green bananas, the dried slices being either used directly for cooking or after grinding into a flour. "Mutere" is used chiefly as a famine reserve and does not feature largely in the diet under normal conditions.

In Gabon, plantains are sometimes made into dried slices which can be stored and used on long journeys, and plantains are used in Cameroon to prepare dried pieces which are stored and ground as needed into flour for use in cooking a paste known as "fufu". Dried green banana slices are also used in parts of South and Central America and West Indies for preparing flour. The other nutritionally important product is beer which is a major product in Uganda, Rwanda and Burundi where green banana utilisation is particularly high.

Preservation Methods and Processes

Drying. - Both ripe and unripe bananas and plantains are normally peeled and sliced before drying, although banana figs are sometimes prepared from whole ripe fruit. Sun drying is the most widespread technique where the climate is suitable but drying in ovens or over fires is also practiced. In west Africa, plantains are often soaked and sometimes parboiled before drying. The slices of unripe fruit are normally spread out on bamboo frameworks; or a cemented area; or on a mat; or on a swept-bare patch of earth; or on a roof; or sometimes on stones outcrops or sheets of corrugated iron. Oven-drying of ripe bananas is practiced in Polynesia as a mean of preserving the fruits, which are then wrapped in leaves and bound tightly to store until needed. In East Africa a method has been reported that involves drying the peeled bananas on a frame placed over a fire for 24 hr before drying in the sun, to accelerate the process.

Product Stability and Storage Problems

There is little experimental data on the storage life of the traditionally made banana and plantain products.

Potential for Scaling up of Traditional Processes to Industrial Level

Many banana products are now produced on an industrial scale, including the traditional banana figs and flour, and the processing techniques are described below. One of the main problems encountered has been the susceptibility of banana products to flavour deterioration and discoloration and in the past many products reaching the market have been of poor quality.

A great deal of research has been directed to overcoming these problems, although however good the resultant products are they cannot compare in flavour and other characteristics with the fresh banana fruit. Indeed, an important constraint on the large-scale development of banana processing is the lack of demand for banana products since the fresh fruit is available throughout the year in most parts of the tropical world.

The production of beer from banana and plantains has not been scaled up to an industrial level, and while an important product in localised areas of tropical Africa, the market is rapidly declining in favour of European-type brews produced locally.

Industrial Processing

The main commercial products made from bananas are canned or frozen purée, dried figs, banana powder, flour, flakes, chips (crisps), canned slices and jams.

Banana products can be divided roughly into two types - those for direct consumption, such as figs, and those for use in food manufacturing industry, for example purées and powder.

Banana figs, or fingers as they are sometimes known, are usually whole, peeled fruit carefully dried so as to retain their shape, although sometimes the fruit is sliced or halved to facilitate drying. Banana and plantain chips (crisps) are thinly sliced pieces of fruit fried in oil and eaten as a snack like potato chips (crisps).

The main use of canned slices is in tropical fruit salads. Banana flakes are used as a flavouring or in breakfast cereals. Banana purée find use mainly in the production of baby foods. Banana flour is said to be highly digestible and is used in baby and invalid foods, but can also be used in the preparation of bread and beverages.

Banana powder is used chiefly in the baking industry for the preparation and fillings for cakes and biscuits and is also used for invalid and baby foods.

Processing Technology

In general, to obtain a good-quality product from ripe-bananas the fruit is harvested green and ripened artificially under controlled conditions at the processing factory. After ripening, the banana hands are washed to remove dirt and any spray residues, and peeled. Peeling is almost always done by hand using stainless steel knives, although a mechanical peeler for ripe bananas has been developed, capable of peeling 450 Kg of fruit per hour (Banana Bulletin, 1974).

The peeling of unripe bananas and plantains is facilitated by immersing the fruit in hot water. For example, immersion in water at 70-75 ° C for 5 min. has been suggested as an aid for peeling green bananas for flour production, while the peeling of green bananas for freezing has been facilitated by immersion in water at 93° C for 30 min.

Banana Figs

Fully ripe fruits with a sugar content of about 19.5% are used and are treated with sulphurous acid after peeling, then dried as soon as possible afterwards. Various drying systems have been described using temperatures between 50 and 82° C for 10 to 24 hr to give a moisture content ranging from 8 to 18% and a yield of dried figs of 12 to 17% of the fresh banana on the stem.

One factory in Australia uses a solar heat collector on the roof to augment the heat used for drying bananas. Bananas can also be dried by osmotic dehydration, using a technique which involves drawing water from 1/4-in. thick banana by placing them in a sugar solution of 67 to 70 deg.

Brix for 8 to 10 hr. followed by vacuum-drying at 65 to 70° C, at a vacuum of 10 mm Hg for 5 hr. The moisture content of the final products is 2.5% or less, much lower than that achieved by other methods.

Banana Purée

Banana purée is obtained by pulping peeled, ripe bananas and then preserving the pulp by one of three methods: canning aseptically, acidification followed by normal canning, or quick-freezing.

The bulk of the world's purée is processed by the aseptic canning technique. Peeled, ripe fruits are conveyed to a pump which forces them through a plate with 1/4-in. holes, then onto a homogeniser, followed by a centrifugal de-aerator, and into a receiving tank with 29in. vacuum, where the removal of air helps prevent discoloration by oxidation.

The purée is then passed through a series of scraped surface heat exchangers where it is sterilised by steam, partially cooled, and finally brought to filling temperature. The sterilised purée is then packed aseptically into steam-sterilised cans which are closed in a steam atmosphere.

Banana Slices

Several methods for canning of banana slices in syrup are used. Best-quality slices are obtained from fruit at an early stage of ripeness. The slices are processed in a syrup of 25 deg. Brix with pH about 4.2, and in some processes calcium chloride (0.2%) or calcium lactate (0.5%) are added as firming agents.

A method for producing an intermediate-moisture banana product for sale in flexible laminate pouches has been developed. Banana slices are blanched and equilibrated in a solution containing glycerol (42.5%), sucrose (14.85%), potassium sorbate (0.45%), and potassium metabisulphite (0.2%) at 90 deg. C for 3 min. to give a moisture content of 30.2%.

Banana Powder

In the manufacture of banana powder, fully ripe banana pulp is converted into a paste by passing through a chopper followed by a colloid mill. A 1 or 2 % sodium metabisulphite solution is added to improve the colour of the final product. Spray- or drum-drying may be used, the latter being favoured as all the solids are recovered.

A typical spray dryer can produce 70 kg powder per hour to give yields of 8 to 11% of the fresh fruit, while drum-drying gives a final yield of about 13% of the fresh fruit. In the latter method the moisture content is reduced to 8 to 12 % and then further decreased to 2 % by drying in a tunnel or cabinet dryer at 60° C.

Banana Flour

Production of flour has been carried out by peeling and slicing green fruit, exposure to sulphur dioxide gas, then drying in a counter-current tunnel dryer for 7 to 8 hr. with an inlet temperature of 75° C and outlet temperature of 45° C, to a moisture content of 8%, and finally milling.

Banana Chips (Crisps)

Typically, unripe peeled bananas are thinly sliced, immersed in a sodium or potassium metabisulphite solution, fried in hydrogenated oil at 180 to 200° C, and dusted with salt and an antioxidant.

Alternatively, slices may be dried before frying and the antioxidant and salt added with the oil. Similar processes for producing plantain chips have been developed.

Banana Beverages

In a typical process, peeled ripe fruit is cut into pieces, blanched for 2 min. in steam, pulped and pectolytic enzyme added at a concentration of 2 g enzyme per 1 kg pulp, then held at 60 to 65° C and 2.7 to 5.5 pH for 30 min.

In a simpler method, lime is used to eliminate the pectin. Calcium oxide (0.5%) is added to the pulp and after standing for 15 min. this is neutralised giving a yield of up to 88% of a clear, attractive juice. In another process banana pulp is acidified, and steam-blanched in a 28-in Hg vacuum which ensures disintegration and enzyme inactivation.

The pulp is then conveyed to a screw press, the resulting purée diluted in the ratio 1:3 with water, and the pH adjusted by further addition of citric acid to 4.2 to 4.3, which yields an attractive drink when this is centrifuged and sweetened.

Jam

A small amount of jam is made commercially by boiling equal quantities of fruit and sugar together with water and lemon juice, lime juice or citric acid, until setting point is reached.

Product Stability and Spoilage Problems

All dried banana products are very hydroscopic and susceptible to flavour deterioration and discoloration, but this can be overcome to some extent by storing in moisture-proof containers and sulphiting the fruit before drying to inactivate the oxidases. The dried products are also liable to attack by insects and moulds if not stored in dry conditions, although disinfestation after drying by heating for 1 hr to 80° C or by fumigation with methyl bromide ensures protection against attack. Banana powder is said to be stored for up to a year commercially and flakes have been stored in vacuum-sealed cans with no deterioration in moisture, colour or flavour for 12 months.

Banana chips tend to have a poor storage life and to become soft and rancid. However, chips treated with an antioxidant have been stored satisfactorily at room temperature in hermetically sealed containers up to 6 months with no development of rancidity.

Quality Control Methods

In general a good quality product is obtained if fruit is harvested at the correct stage of maturity and, where appropriate, ripened under controlled conditions. For example, in the case of banana figs, the fruit should be fully mature (sugar content of 19.5% or above) or the final product is liable to be tough and lacking in flavour. However, if over-ripe fruit is used, the figs tend to be sticky and dark in colour, so the fruit must be fully yellow but still firm.

For banana flour, which is prepared from unripe bananas, the fruit is harvested at three-quarters the full-ripe stage and is processed within 24 hr. prior to the onset of ripening. If less mature fruit is used, the flour tastes slightly astringent and bitter due to the tannin content. Bananas harvested between 85 and 95 days after the emergence of the inflorescence, with a pulp-to-peel ratio of about 1.7, were considered to be most suitable for the deep-fat frying.

Other criteria suggested for assessing maturity were beta-carotene and reducing sugar content, both of which increase with increasing maturity and pH which decreases as the fruit ripens, and these should be, respectively, about 2000 μg/100 g, less than 1.5% and 5.8 or above. Browning was found to occur if the sugar content was higher than 1.5%. The determination of crude fat in processed chips is also considered to be a necessary quality control measure.

It is important to remove all impurities prior to processing of products, and this is done by washing to remove dirt and spray residues and control on the processing line so that substandard fruit can be removed.

Preparation Methods for Fresh Bananas and Plantains

The main ways of preparing fresh bananas for consumption are boiling or steaming, roasting or baking and frying. Boiling followed by pounding into "fufu" is also widely adopted in certain areas of the tropics.

Boiling or Steaming

Plantains and bananas are often prepared simply by boiling in water, either in their peel or after peeling, and either ripe or unripe; if unripe, the fruit is scraped thoroughly after peeling to remove all traces of fibrous material. The boiled fruit is eaten alone or more usually accompanied by a sauce. This preparation technique is widely used in West Africa.

Roasting or baking. Unpeeled or peeled fruit, either ripe or unripe, is roasted simply by placing in the ashes of a fire or in an oven. This method is widely used in West Africa, East Africa and the South Pacific islands. For example, ripe plantains are placed unpeeled in an oven and when partly brown and tender, removed and peeled, then replaced in the oven and roasted evenly.

Frying. Ripe or unripe plantains or bananas are often peeled, sliced and cooked in oil, particularly in West Africa and in parts of South America and the West Indies. Similar products are also made in East Africa. Typically, ripe plantains are peeled, cut into slices or split lengthways, and fried in palm oil or with groundnut oil, the pieces being served either hot with a sauce or with fried eggs, or cold as a snack.

Pounding. Pounding is a process, used particularly in West Africa, for preparing most perishable staple food crops including plantains, cassava, yams and cocoyams to obtain a paste or dough known as "fufu" (also spelled "foofoo", "foutou", "foufou"). The plantains are peeled or boiled and peeled after boiling and pounded in a wooden mortar, the resulting paste normally being eaten with soup or a spiced sauce of meat and vegetables, but sometimes after wrapping in leaves and steaming.

Mango and Guava Processing Technologies

Mangoes are processed at two stages of maturity. Green fruit is used to make chutney, pickles, curries and dehydrated products. The green fruit should be freshly picked from the tree. Fruit that is bruised, damaged, or that has prematurely fallen to the ground should not be used. Ripe mangoes are processed as canned and frozen slices, purée, juices, nectar and various dried products. Mangoes are processed into many other products for home use and by cottage industry. The mango processing presents many problems as far as industrialization and market expansion is concerned. The trees are alternate bearing and the fruit has a short storage life; these factors make it difficult to process the crop in a continuous and regular way. The large number of varieties with their various attributes and deficiencies affects the quality and uniformity of processed products.

The lack of simple, reliable methods for determining the stage of maturity of varieties for processing also affects the quality of the finished products. Many of the processed products require peeled or peeled and sliced fruit. The lack of mechanised equipment for the peeling of ripe mangoes is a serious bottleneck for increasing the production of these products.

A significant problem in developing mechanised equipment is the large number of varieties available and their different sizes and shapes. The cost of processed mango products is also too expensive for the general population in the areas where most mangoes are grown. There is, however, a considerable export potential to developed countries but in these countries the processed mango products must compete with established processed fruits of high quality and relatively low cost.

Green Mango Processing

Pickles. The optimum stage of maturity should be determined for each variety used to make pickles.

There are two classifications of pickles - salt pickles and oil pickles. They are processed from whole and sliced fruit with and without stones. Salt is used in most pickles.

The many kinds of pickles vary mainly in the proportions and kinds of spices used in their preparation. One basic recipe for the study of the preparation and storage of pickles in oil is as follows:

Mango pieces	250 g	Tumeric powder	2 to 4 g
Salt	60 g	Fenugreek seeds	2 to 4 g
Mustard powder	30 g	Bengal gram seeds	2 to 4 g
Chili powder	20 g	Gingelly oil	20 to 30 g

The ingredients are mixed together and filled into wide-mouthed bottles of 0.5 kg capacity. Three days later the contents are thoroughly mixed and refilled into the bottles. Extra oil is added to form a 1-2 cm layer over the pickles.

Chutney. The product is prepared from peeled, sliced or grated unripe or semi-ripe fruit by cooking the shredded fruit with salt over medium heat for 5 to 7 minutes, mixed and then sugar, spices and vinegar are added. Cook over moderate heat until the product resembles a thick purée, add remaining ingredients and simmer another 5 min. Cool and preserve in sterilised jars.

Spices usually include cumin seeds, ground cloves, cinnamon, chili powder, ginger and nutmeg. Other ingredients such as dried fruits, onions, garlic and nuts may be added.

Drying/dehydration. Immature fruit is peeled and sliced for sun-drying. The dried mango slices can be powdered to make a product called amchoo. The use of blanching, sulphuring and mechanical dehydration gives a product with better colour, nutrition, storability and fewer microbiological problems.

Ripe Mango Processing

Purée. Mangoes are processed into purée for re-manufacturing into products such as nectar, juice, squash, jam, jelly and dehydrated products. The purée can be preserved by chemical means, or frozen, or canned and stored in barrels. This allows a supply of raw materials during the remainder of the year when fresh mangoes are not available.

It also provides a more economical means of storage compared with the cost of storing the finished products, except for those which are dehydrated, and provides for more orderly processing during peak availability of fresh mangoes.

Mangoes can be processed into purée from whole or peeled fruit. Because of the time and cost of peeling, this step is best avoided but with some varieties it may be necessary to avoid off-flavours which may be present in the skin. The most common way of removing the skin is hand-peeling with knives but this is time-consuming and expensive. Steam and lyepeeling have been accomplished for some varieties.

Several methods have been devised to remove the pulp from the fresh ripe mangoes without hand-peeling. A simplified method is as follows: the whole mangoes were exposed to atmospheric steam for 2 to 2 1/2 min in a loosely covered chamber, then transferred to a stainless steel tank.

The steam-softened skins allowed the fruit to be pulped by a power stirrer fitted with a saw-toothed propeller blade mounted 12.7 to 15.2 cm below a regular propeller blade. The pulp is removed from the seeds by a continuous centrifuge designed for use in passion fruit extraction. The pulp material is then passed through a paddle pulper fitted with a 0.084 cm screen to remove fibre and small pieces of pulp.

Mango purée can be frozen, canned or stored in barrels for later processing. In all these cases, heating is necessary to preserve the quality of the mango purée. In one process, purée is pumped through a plate heat exchanger and heated to 90°C for 1 min and cooled to 35° C before being filled into 30 lb tins with polyethylene liners and frozen at -23.50 C.

In an other process, pulp is acidified to pH 3.5, pasteurized at 90°C, and hot-filled into 6 kg high-density bulk polyethylene containers that have been previously sterilised with boiling water. The containers are then sealed and cooled in water. This makes it possible to avoid the high cost of cans.

Wooden barrels may be used to store mango pulp in the manufacture of jams and squashes. The pulp is acidified with 0.5 to 1.0% citric acid, heated to boiling, cooled, and SO2 is added at a level of 1000 to 1500 ppm in the pulp. The pulp is then filled into barrels for future use.

Slices. Mango slices can be preserved by canning or freezing, and recent studies have shown the feasibility of pasteurized-refrigerated and dehydro-canned slices. The quality of the processed product in all of these procedures will be dependent upon selection of a suitable variety along with good processing procedures. Thermal process canning of mango slices in syrup is the most widely used preservation method.

Beverages : The commercial beverages are juice, nectar and squash. Mango nectar and juice contain mango purée, sugar, water and citric acid in various proportions depending on local taste, government standards of identity, pH control, and fruit composition of the variety used. Mango squash in addition to the above may contain SO2 or sodium benzoate as a preservative. Other food grade additives such as ascorbic acid, food colouring, or thickeners may be used in mango beverages.

A short description of finished products found in literature is as follows:

- mango juice: prepared by mixing equal quantities of pulp (purée) and water together and adjusting the total soluble solids (TSS) and acidity to taste (12 to 15% TSS and 0.4 to 0.5% acidity as citric acid);
- mango nectar containing 25% purée can be prepared using the following procedure.

		Brix of purée	
Nectar components	15°	17°	20°
Purée	100	100	100
Sugar	45	43	40
Water	255	257	260

Commercial processing conditions may require the use of a preservative.

The pH is adjusted to approximately 3.5 by adding citric acid as a 50% solution.

The time of heat processing will vary with filling temperature, can size and viscosity of the juice or nectar. Mango squash may be prepared according to flow-sheet described below; the finished product may contain 25% juice, 45% TSS and 1.2 to 1.5% acidity and may be preserved with sulphur dioxide (350ppm) or sodium benzoate (1000 ppm) in glass bottles.

Mango squash simplified flow-sheet.

Table: *Ingredients*

Mango pulp	900	900
Sugar	900	1100
Citric acid	18	15
Water	900	900

Mangoes are washed, stored, peeled with stainless steel knives. The pulp is prepared by using a pulper with fine sieve (0.025-in); Sugar is mixed with water and citric acid = syrup; The pulp is added to the syrup and mixed well; The mixture is strained trough cloth; The squash is heated at 85° C and bottles are filled and closed.

For additional heat treatment bottles may need to be maintained at a product temperature of 80°C for 30 minutes if the product is to be processed without preservatives. The bottles are then left to cool in water and stored at room temperature.

Two negative points must be avoided: presence of air bubbles (which is a source of quick deterioration) and separation of squash solids (giving an undesirable appearance). The means to avoid these two phenomena are described in the fruit juices section.

A type of "squash type" beverage may also be manufactured with '/a pulp + '/a water + i/a sugar and pH adjusted to 3.7 by addition of citric acid. Using different sieve sizes affects the quality and reduces air bubbles to a certain extent but homogenisation and de-aeration of purée or squash seem to be important in order to avoid separation and air bubbles.

The squash quality is evaluated on the basis of the following characteristics: pH, titrable acidity, soluble solids, ascorbic acid (by 2,6 dichlorophenol indophenol method), specific gravity.

Dried/dehydrated. Ripe mangoes are dried in the form of pieces, powders, and flakes. Drying procedures such as sun-drying, tunnel dehydration, vacuum-drying, osmotic dehydration may be used. Packaged and stored properly, dried mango products are stable and nutritious.

One described process involves as pre-treatment dipping mango slices for 18 hr (ratio 1:1) in a solution containing 40° Brix sugar, 3000 ppm SO2, 0.2% ascorbic acid and 1% citric acid; this method is described as producing the best dehydrated product. Drying is described using an electric cabinet through flow dryer operated at 60° C. The product showed no browning after 1 year of storage.

Drum-drying of mango purée is described as an efficient, economical process for producing dried mango powder and flakes. Its major drawback is that the severity of heat preprocessing can produce undesirable cooked flavours and aromas in the dried product. The drum-dried products are also extremely hydroscopic and the use of in-package desiccant is recommended during storage.

Canning. This preservation technology is described in various technological flow-sheets in this bulletin.

Mango bar or "fruit leather" is presented in various flow-sheets.

Guava Processing Technologies

Guava purée. Guava purée is used in the manufacture of guava nectar, various juice drink blends and in the preparation of guava jam. The washed sound fruit is first passed through a chopper or slicer to break up the fruit and this material is fed into a pulper. The pulper will remove the seeds and fibrous pieces of tissue and force the reminder of the product through a perforated stainless steel screen. The holes in the screen should be between 0.033 and 0.045 in. The machine should be fed at a constant rate to ensure efficient operation.

The puréed material coming from the pulper is next passed through a finisher. The finisher is equipped with a screen containing holes of approximately 0.020 in. The finisher will remove the stone cells from the fruit and provide the optimum consistency to the product.

Perhaps the best way to preserve the guava purée is by freezing and the material passing through the finisher can be packaged and frozen with no further treatment. It is not necessary to heat the product to inactivate enzymes or for other purposes. The material can be frozen in a number of types of cartons and cans; however, a fibre box with a plastic bag inside is commonly used and is probably the less expensive.

It is also possible to can and heat process the guava purée and this can be accomplished by heating the purée to 195° F in an open double bottom kettle, filling into cans, closing the cans, inverting the cans for a few seconds, followed by cooling. Cans should be cooled rapidly to approximately 100° F before they are cased and stacked into warehouses.

Guava Juice and Concentrate

Guava juice can be used in the manufacture of a clear guava jelly or in various drinks. A clear juice may be prepared from guava purée that is depectinised enzymatically. About 0.1% pectin-degrading enzyme is mixed into the purée at room temperature; heating of the product at approximately 120° F will greatly speed the action of the enzyme. After 1 hr. clear juice is separated from the red pulp by centrifuging or by pressing in a hydraulic juice press. A batch-type or continuous-flow centrifuge can be used on the depectinised purée with no further treatment.

The clear juice after centrifuge or after press (and subsequent filtration) can be preserved by freezing or by pasteurization in hermetically sealed cans.

For shipment to overseas markets it may be advantageous to concentrate either the purée or the juice.

Recent Trends in Fruit and Vegetable Processing

New products. The number and variety of fruit and vegetable products available to the consumer has increased substantially in recent years. The fruit and vegetable industry has undoubtedly benefited from the increased recognition and emphasis on the importance of these products in a healthy diet.

Traditional processing and preservation technologies such as heating, freezing and drying together with the more recent commercial introduction of chilling continue to provide the consumer with increased choice.

This has been achieved by new heating (e.g. UHT, microwave, ohmic) and freezing (e.g. cryogenic) techniques combined with new packaging materials and technologies (e.g. aseptic, modified atmosphere packaging).

The overall trend in new fruit and vegetable products is "added value", thus providing increased convenience to the consumer by having much greater variety of ready prepared fruit and vegetable products. These may comprise complete meals or individual components. The suitability of products and packages for microwave re-heating has been an important factor with respect to added convenience.

Table: *Numbers of new fruit and vegetable products*

	1990	*1991*	*1992 Jan-June*
Frozen vegetable products	66	95	21
Chilled vegetable products	76	81	78
Heat processed vegetables	51	60	38
Heat processed fruits	13	14	5
Fruit juices and drinks	73	83	46
Potato crisps	32	33	16

Source: C. Dennis (1993)

New product development in the fruit and vegetable sector is most important in meeting the continued challenge of providing the consumer with choice and high quality products.

A Fresh Look at Dried Fruit

New fruit varieties and advance in drying technologies are putting a fresh twist on dried fruit applications. Fruits that have been introduced to the drying process include cranberries, blueberries, cherries, apples, raspberries and strawberries - not to mention the traditional mainstays of raisins, dates, apricots, peaches, prunes and figs.

Perceived as a "value-added" ingredient, dried fruit adds flavour, colour, texture and diversity with little alteration to an existing formula. The growing interest in ethnic cuisines in U.S.A. and the change to a more healthy way of eating, has also moved dried fruit considerably closer to the mainstream.

Found primarily in the baking industry, dried fruit is coming into its own in various food products, including entrees, side dishes and condiments. Compotes, chutneys, rice and grain dishes, stuffings, sauces, breads, muffins, cookies, deserts, cereals and snacks are all food categories encompassing dried fruit.

Since some dried fruit is sugar infused (osmotic drying), food processors can decrease the amount of sugar in formula - this is especially the case in baked products. Processors are making adjustments in moisture content of the dried fruit so that a varied range is available for different applications. An added bonus is dried fruits' shelf stability (a shelf life of at least 12 months). Dried fruit is more widely available in different forms, including whole dried, cut, diced and powders.

Citric acid and its use in fruit and vegetable processing. Citric acid may be considered as "Nature's acidulant". It is found in the tissues of almost all plants and animals, as well as many yeasts and moulds.

Commercially citric acid is manufactured under controlled fermentation conditions that produce citric acid as a metabolic intermediate from naturally-occurring yeasts, moulds and nutrients. The recovery process of citric acid is through crystallization from aqueous solutions.

Citric acid is widely used in carbonated and still beverages, to impart a fresh-fruit "tanginess". Citric acid provides uniform acidity, and its light fruity character blends well and enhances fruit juices, resulting in improved palatability. The amount of citric acid used depends on the particular desired flavour (e.g., High-acid: lemonade; Medium-acid: orange, punch, cherry; Low-acid: strawberry, black cherry, grape).

Sodium citrate is often added to beverages to mellow the tart taste of high acid concentrations. It provides a cool, distinctive smooth taste and masks any bitter aftertaste of artificial sweeteners. In addition, it serves as a buffer to stabilise the pH at the desired level. The high water solubility of citric acid (181 g/100 ml) makes it an ideal additive for fountain fruit syrups and beverages concentrates as a flavour enhancer and microbial growth inhibitor (preferably at pH < 4.6).

In processed fruits and vegetables, citric acid performs the following functions:

a. It reduces heat-processing requirements by lowering pH: inhibition of microbial growth is a function of pH and heat treatment. Higher heat exposure and lower pH result in greater inhibition. Thus the use of citric acid to bring pH below 4.6 can reduce the heating requirements. In canned vegetables, citric acid usage is greatest in tomatoes, onions and pimentos. For tomato packs, the National Canners Association recommends a pH of 4.1 to 4.3. In general, 0. 1% citric acid will reduce the pH of canned tomatoes by 0.2 pH units.

b. Optimise flavour: citric acid is added to canned fruits to provide for adequate tartness. Recommended usage level is generally less then 0.15%.

c. Supplement antioxidant potential: citric acid is used in conjunction with antioxidants such as ascorbic and erythorbic acids, to inhibit colour and flavour deterioration caused by metal-catalysed enzymatic oxidation. Recommended usage levels are generally 0.1% to 0.3% with the antioxidant at 100 to 200 ppm.

d. Inactivate undesirable enzymes: oxidative browning in most fruits and vegetables is catalysed by the naturally present polyphenol oxidase. The enzymatic activity is strongly dependent on pH.

Addition of citric acid to reduce pH below 3 will result in inactivation of this enzyme and prevention of browning reactions.

Cherry and Apricot Oils are Safe for Food Use

The oils obtained by cold pressing the kernels of the cherry (Prunus cerasus) and the apricot (Prunus armeniaca) have been declared acceptable for food use by the UK Ministry of Agriculture, Fisheries & Food's Advisory Committee on Novel Foods and Processes (ACNFP) by June 1993.

In its assessments of the safety in use of the cherry and apricot kernel oils, the ACNFP consider specifications that included data on fatty acid composition, the presence of natural antioxidants and the content of cyanide, mycotoxins and heavy metals.

The Committee says that it gave particular consideration to the possible presence in the oils of the cyanogenic glucoside amygdalin, from which cyanide is released by enzymic action when the kernels of cherry and apricot are crushed. Amygdalin was found to be absent from the cherry and apricot kernel oils.

The ACNFP is satisfied that there are no food safety reasons why the use of cherry and apricot kernel should not be acceptable provided there is compliance with the specifications shown in Table below.

Table: *Specification of purity for cherry and apricot kernel oils as determined by UK ACNFP*

	Cherry	*Apricot*
Contaminants limits:		
Heavy metals (total)	0.5 mg/kg	0.5 mg/kg
Aflatoxins (total)	4.0 g/kg	4.0 g/kg
Cyanide	0.15 mg/kg	0.15 mg/kg
Pesticide residues	0.01 mg/kg	0.01 mg/kg
Tocopherols:		
alpha/delta/gamma (mg/kg)	356-886	569-899

Source: Anon. (1993)

The oils are obtained by the mechanical mincing and cold pressing of kernels extracted from cleaned cherry or apricot stones. After filtering, the oils are stored and are to be sold in a raw, unrefined state. The cherry and apricot kernel oils are high unsaturated and are expected to be used as speciality oils for salad dressings, baking and shallow frying applications.

The Use of Fruit Juices in Confectionery Products

During the last decade, the concept of fruit juices has gained immensely on consumer popularity. The majority of new non-alcoholic and alcoholic fruit drink products were a combination of syrups, fruit juices and flavours.

The confectionery industry followed suit and new products incorporated fruit juices as part of their confectionery formulations and processes. Fruit juice concentrates of high solids are often used instead of normal or single-fold juices.

Juice concentrates are made of pure fruit juices. The process starts with pressing fruits and obtaining pure fruit juice; this is stabilised by heat treatment which inactivates enzymes and micro-organisms. The next processing step is concentration under vacuum up to 40-65° Brix or 4-7 fold. The concentrates are then blended for standardisation and stored.

These fruit juice concentrates are often further stabilised by the addition of sodium benzoate and potassium sorbate and are usually stored away from light and are refrigerated or frozen.

Depectinised fruit juices are also used to prevent foaming in confectionery processes and are essential for use in clear beverage products. Fruit juice concentrates which are depectinised, and have added preservatives are called stabilised, clarified, fruit juice concentrates.

Fruit juices are used in confectionery products in conjunction with natural and artificial flavours which provides intense flavour impact and are cost-effective for a confectionery product.

The traditional concern in using fruit juice concentrates in confectionery applications has been the effect of the natural acids on the finished product, particularly the formation of invert sugar during processing.

This is a logical concern since concentrates contain differing amounts and types of acids. For example: apple, cherry, strawberry and other berries contain primarily malic acid. Grapes mainly contain tartaric acid. Cranberry is high in quinic acid. Citrus fruits and pineapple contain differing amounts of citric acid. The concentrates, when used, are normally buffered to a pH of 5-7 with sodium hydroxide.

In formulating products with fruit juice concentrates, the solids of the concentrate are considered as mostly reducing sugars and a reduction in corn syrup is made to compensate for equivalent amount of reducing sugar being added in the concentrate.

The exact replacement can be determined by measuring the D.E. of the concentrate to be added. In formulations when small amounts of concentrate are used (less than 1%), no adjustment is made since the reducing sugar contribution of the concentrate is not significant.

Fruit juice concentrates can also be used to provide a source of natural colour, in particular red colour. Grape, raspberry, cherry, strawberry and cranberry concentrates in small amounts are very effective in colouring cream centres.

The inclusion of fruit juices in confectionery products is now left up to the imagination of the manufacturer. These products must, of course, hold up to the standards of flavour integrity, and product excellence, during the shelf-life of these products.

Vegetable Specific Processing Technologies

Vegetable processors must appreciate the substantial differences that varieties of a given vegetable will possess. In addition to variety and genetic strain differences with respect to weather, insect and disease resistance, varieties of a given vegetable will differ in size, shape, time of maturity, and resistance to physical damage.

Varietal differences then further extend into warehouse storage stability, and suitability for such processing methods as canning, freezing, pickling or drying. A variety of peas that is suitable for canning may be quite unsatisfactory for freezing and varieties of potatoes that are preferred for freezing may be less satisfactory for drying or potato chip manufacture. This should be expected since different varieties of a given vegetable will vary somewhat in chemical composition, cellular structure and biological activity of their enzyme system.

Harvesting and Pre-processing

When vegetables are maturing in the field they are changing from day to day. There is a time when the vegetable will be at peak quality from the stand-point of colour, texture and flavour.

This peak quality is quick in passing and may last only a day. Harvesting and processing of several vegetables, including tomatoes, corn and peas are rigidly scheduled to capture this peak quality.

After the vegetable is harvested it may quickly pass beyond the peak quality condition. This is independent of microbiological spoilage; these main deteriorations are related to:

a) loss of sugars due to their consumption during respiration or their conversion to starch; losses are slower under refrigeration but there is still a great change in vegetable sweetness and freshness of flavour within 2 or 3 days;

b) production of heat when large stockpiles of vegetables are transported or held prior to processing.

   At room temperature some vegetables will liberate heat at a rate of 127,000 kJ/ton/day; this is enough for each ton of vegetables to melt 363 kg of ice per day. Since the heat further deteriorates the vegetables and speeds micro-organisms growth, the harvested vegetables must be cooled if not processed immediately.

   But cooling only slows down the rate of deterioration, it does not prevent it, and vegetables differ in their resistance to cold storage. Each vegetable has its optimum cold storage temperature which may be between about 0-100 C (32-500 F).

c) the continual loss of water by harvested vegetables due to transpiration, respiration and physical drying of cut surfaces results in wilting of leafy vegetables, loss of plumpness of fleshy vegetables and loss of weight of both.

Moisture loss cannot be completely and effectively prevented by hermetic packaging. This was tried with plastic bags for fresh vegetables in supermarkets but the bags became moisture fogged, and deterioration of certain vegetables was accelerated because of buildup of CO2 and decrease of oxygen in the package. It therefore is common to perforate such bags to prevent these defects as well as to minimise high humidity in the package which would encourage microbial growth.

Shippers of fresh vegetables and vegetable processors, whether they can, freeze, dehydrate, or manufacture soups or ketchup, appreciate the instability and perishability of vegetables and so do everything they can to minimise delays in processing of the fresh product. In many processing plants it is common practice to process vegetables immediately as they are received from the field.

To ensure a steady supply of top quality produce during the harvesting period the large food processors will employ trained field men; they will advise on growing practices and on spacing of plantings so that vegetables will mature and can be harvested in rhythm with the processing plant capabilities. This minimises stockpiling and need for storage.

Cooling of vegetables in the field is common practice in some areas. Liquid nitrogen-cooled trucks may next provide transportation of fresh produce to the processing plant or directly to market.

Upon arrival of vegetables at the processing centre the usual operations of cleaning, grading, peeling, cutting and the like are performed using a moderate amount of equipment but a good deal of hand labour also still remains.

Reception. This covers qualitative and quantitative control of delivered vegetables. The organoleptic control and the evaluation of the sanitary state, even if they are very important steps in vegetables' characteristics assessment, cannot establish their technological value.

On the other hand, laboratory controls do not precisely establish their technological properties because of the difficulty in putting into showing some deterioration when using rapid control methods.

One correct method of vegetable quality appraisal is their overall evaluation based on the whole complex of data that can be obtained by combining an extensive organoleptic evaluation with simple analysis that can be performed rapidly in plant laboratory. These analysis can be:

a. refractometric extract (tomatoes, fruit, etc.);

b. specific weight (potatoes, peas, etc.);

c. consistency (measured with tenderometers, penetrometers, etc.);

d. boiling tests, etc.

Temporary storage. This step should be as short as possible and better completely eliminated. Vegetables can be stored in:

a. simple stores, without artificial cooling;

b. in refrigerated stores; or, in some cases,

c. in silos (potatoes, etc.).

Simple stores should be covered, fairly cool, dry and well ventilated but without forced air circulation which can induce significant losses in weight through intensive water evaporation; air relative humidity should be at about 70-80%.

Refrigerated storage is always preferable and in all cases a processing centre needs a cold room for this purpose, adapted in volume I capacity to the types and quantities of vegetables (and fruits) that are further processed.

Washing. Washing is used not only to remove field soil and surface micro-organisms but also to remove fungicides, insecticides and other pesticides, since there are laws specifying maximum levels of these materials that may be retained on the vegetable; and in most cases the allowable residual level is virtually zero. Washing water contains detergents or other sanitisers that can essentially completely remove these residues.

The washing equipment, like all equipment subsequently used, will depend upon the size, shape and fragility of the particular kind of vegetable:

- flotation cleaner for peas and other small vegetables;
- rotary washer in which vegetables are tumbled while they are sprayed with jets of water; this type of washer should not be used to clean fragile vegetables;

Sorting. This step covers two separate operations:

a) removal of non-standard vegetables (and fruit) and possible foreign bodies remaining after washing;

b) quality grading based on variety, dimensional, organoleptical and maturity stage criterion.

Skin Removal/peeling. Some vegetables require skin removal. This can be done in various ways.

Mechanical : This type of operation is performed with various types of equipment which depend upon the result expected and the characteristics of the fruit and vegetables, for example:

i. a machine with abrasion device (potatoes, root vegetables);

ii. equipment with knives (apples, pears, potatoes, etc.);

iii. equipment with rotating sieve drums (root vegetables). Sometimes this operation is simultaneous with washing (potatoes, carrots, etc.) or preceded by blanching (carrots).

Chemical: Skins can be softened from the underlying tissues by submerging vegetables in hot alkali solution. Lye may be used at a concentration of about 0.5-3%, at about 93° C (2000 F) for a short time period (0.5-3 min). The vegetables with loosened skins are then conveyed under high velocity jets of water which wash away the skins and residual lye.

In order to avoid enzymatic browning, this chemical peeling is followed by a short boiling in water or an immersion in diluted citric acid solutions. It is more difficult to peel potatoes with this method because it is necessary to dissolve the cutin and this requires more concentrated lye solutions, up to 10%.

Thermal: Wet heat (steam). Other vegetables with thick skins such as beets, potatoes, carrots and sweet potatoes may be peeled with steam under pressure (about 10 at) as they pass through cylindrical vessels. This softens the skin and the underlying tissue. When the pressure is suddenly released, steam under the skin expands and causes the skin to puff and crack. The skins are then washed away with jets of water at high pressure (up to 12 at).

Dry heat (flame). Other vegetables such as onions and peppers are best skinned by exposing them to direct flame (about 1 min at 1000° C) or to hot gases in rotary tube flame peelers. Here too, heat causes steam to develop under skins and puff them so that they can be washed away with water.

Manual peeling only use when the other methods are impossible or sometimes as a completion of the other three ways. Average losses at this step are given in Table below.

Table: *Losses at vegetable peeling, in %*

	Peeling methods		
Vegetables	**Manual**	**Mechanical**	**Chemical**
Potatoes	15-19	18-28	-
Carrots	13-15	16-18	8-10
Beets	1416	13-15	9-10

Size reduction. This step is applied according to specific vegetable and processing technology requirements.

Blanching. The special heat treatment to inactivate enzymes is known as blanching. Blanching is not indiscriminate heating. Too little is ineffective, and too much damages the vegetables by excessive cooking, especially where the fresh character of the vegetable is subsequently to be preserved by processing. This heat treatment is applied according to and depends upon the specificity of vegetables, the objectives that are followed and the subsequent processing/ preservation methods.

Two of the more heat resistant enzymes important in vegetables are catalase and peroxidase. If these are destroyed then the other significant enzymes in vegetables also will have been inactivated. The heat treatment to destroy catalase and peroxidase in different vegetables are known, and sensitive chemical tests have been developed to detect the amounts of these enzymes that might survive a blanching treatment. Catalase and peroxidase inactivation tests are presented in section 9.2.9.

Because various types of vegetables differ in size, shape, heat conductivity, and the natural levels of their enzymes, blanching treatments have to be established on an experimental basis. As with sterilisation of foods in cans, the larger the food item the longer it takes for heat to reach the centre. Small vegetables may be adequately blanched in boiling water in a minute or two, large vegetables may require several minutes.

Blanching as a unit operation is a short time heating in water at temperatures of 100° C or below. Water blanching may be performed in double bottom kettles, in special baths with conveyor belts or in modern continuous blanching equipment.

In order to reduce losses of hydrosoluble substances (mineral salts, vitamins, sugars, etc.) occurring during water blanching, several methods have been developed:

- temperature setting at 85-95° C instead of 100° C;
- blanching time has to be just sufficient to inactivate enzymes catalase and peroxidase;
- assure elimination of air from tissues.

An illustration of blanching parameters is seen in Table below.

***Table:** Blanching parameters for some vegetables*

Vegetables	*Temperature, °C*	*Time, min.*
Peas	85-90	2-7
Green beans	90-95	2-5
Cauliflower	Boiling	2
Carrots	90	3-5
Peppers	90	3

Steam heat treatment can also be applied instead of water blanching as a preliminary step before freezing or drying, as long as the preservation method is only used for enzyme inactivation and not to modify consistency.

For drying, the vegetables are conveyed directly from steaming equipment to drying installations without cooling. Vegetable steaming is carried out in continuous installations with conveyer belts made from metallic sieves.

Cooling of vegetables after water blanching or steaming is performed in order to avoid excessive softening of the tissues and has to follow immediately after these operations; one exception is the case of vegetables for drying which can be transferred directly to drying equipment without cooling.

Natural cooling is not recommended because is too long and generates significant losses in vitamin C content. Cooling in pre-cooled air (from special installations) is sometimes used for vegetables that will be frozen.

Cooling in water can be achieved by sprays or by immersion; in any case the vegetables have to reach a temperature value under 37° C as soon as possible. Too long a cooling time generates supplementary losses in valuable hydrosoluble substances; in order to avoid this, the temperature of the cooling water has to be as low as possible.

Canning: Large quantities of vegetable products are canned. A typical flow sheet for a vegetable canning operation (which also applies to fruit for the most part) covers some food process unit operations performed in sequence: harvesting; receiving; washing; grading; heat blanching; peeling and coring; can filling; removal of air under vacuum; sealing/closing, retorting/heat treatment; cooling; labelling and packing. The vegetable may be canned whole, diced, puréed, as juice and so on.

On-line Simplified Methods for Enzyme Activity Check

Peroxidase test:

Solutions. In order to check the peroxidase activity two solutions have to be prepared:

- 1% guaiacol in alcohol solution (1 g guaiacol is dissolved in about 50 cm^3 of 96% ethylic alcohol and then this preparation is brought to 100 cm° with the same solvent);
- peroxide solution 0.3% (1 cm^3 perhydrol is brought to 100 cm^3 with distilled water.

Sampling. From various parts of the material samples are taken (about 20-30 pieces, etc.); the material is then crushed in a laboratory bowl in order to obtain an average sample.

Check. Prom the average sample, 10-20 g of material is taken in a medium capacity test tube; on this sample are poured: 20 cm^3 distilled water; 1 cm^3 of 1% guaiacol solution; 1.6 cm^3 of peroxide solution.

The contents of the test tube is shaken well. The gradual appearance of a weak pink colour indicates an incomplete peroxidase inactivation - reaction slightly positive. If there are no tissue colour modifications after 5 minutes, the reaction is negative and the enzymes have been inactivated.

As an orientative check it is also possible to simply pour a few drops of 1% guaiacol solution and 0.3% peroxide solution directly on blanched and crushed vegetables. A rapid and intensive brown-reddish tissue colouring indicates a high peroxidase activity (positive reaction).

Catalase Test

In order to identify the catalase enzyme activity, 2 g of dehydrated vegetables are well crushed and mixed with about 20 cm^3 of distilled water. After 15 min softening, 0.5 cm^3 of a 0.5% or 1% peroxide solution is poured on prepared vegetables. In the presence of catalase, a strong oxygen generation is observed for about 2-3 minutes.

These tests are of a paramount importance in order to determine the vegetable blanching treatments (temperature and time); incomplete enzyme inactivation has a negative effect on finished product quality.

For cabbage catalase inactivation by blanching is sufficient; blanching further to peroxidase inactivation would have negative effects on product quality and even complete browning.

For all other vegetables and for potatoes, both tests MUST be negative, for catalase and for peroxidase. Quality control/quality assurance and international trade; good manufacturing practices (gmp); hygiene requirements; hazard analysis and critical control points (haccp).

Quality Control/Quality Assurance and International Trade

The international trade in processed fruits and vegetables is very large with an ever increasing number of different types being processed and exported. Whereas once, processing was limited to mostly temperate climate fruits and vegetables, the change has now broadened to include tropical and subtropical types.

The reasons are twofold. Firstly, consumers' dietary habits have become more diverse so that, for example people living in North America may very well like fruit and vegetables grown in Africa or Asia. Secondly, processing techniques, whether they be for canning, freezing or drying, have been improved to an extent where final product is palatable, nutritious and of long and reliable shelf life.

Many developing countries have taken advantage of the continuing worldwide demand for processed fruits and vegetables and earned valuable foreign exchange from exports of products to profitable markets.

The export quality control and inspection of processed fruits and vegetables is directed at ensuring that the final products: have been processed in a registered export establishment that is constructed, equipped and operated in an hygienic and efficient manner;

conform to the requirements of the export regulations for processed fruits and vegetables, and those of the importing country, in respect of such things as quality grades, defects, ingredients, packaging materials, styles, additives, contaminants, fill of container, drained weight; and, conform to labelling requirements.

Inspection and Certification Procedures

In most countries, in processing fruits and vegetables for export, it is not customary to apply continuous inspection as it is in the case of meat. Few, if any, importing countries require it, and the nature of the products themselves is such that only part time check inspection is required during processing together with statistically based inspection, including sampling and analysis, of final product.

However, in circumstances where an establishment is processing export product for the first time, it can be argued that there is an advantage in adopting continuous inspection until the operation is satisfactorily established.

In any event, inspection of raw materials should be carried out at the commencement of each processing run to ensure that only sound fruit or vegetables of sufficient maturity (degree of ripeness) are used for processing. Sampling checks of raw materials should be carried out as frequently as the inspector thinks necessary.

The inspector must ensure that adequate hygiene practices are followed during the processing of the product. For example, in the case of canned and frozen products and other processing methods, raw materials should be washed absolutely clean so that fruit and vegetables entering the processing line are free from dirt, superficial residues of agricultural chemicals, insects and extraneous plant material.

In the case of dried product, especially where the raw material is sun-dried on drying greens or racks, care must be taken to minimize contamination by bird and animal droppings, dust and extraneous plant material. It is often necessary to wash the dried product to ensure cleanliness of the final product.

In the case of canning and freezing, the inspector must obtain full details of the processing programme for at least the following day from management, so that an adequate inspection programme can be scheduled.

In much the same way as for fresh fruit and vegetables, the inspector must also be aware of the pesticides and other chemicals used in the production of the raw materials. Necessary laboratory analyses can then be arranged to ensure residue levels in the final product do not exceed tolerances adopted by importing countries.

At the commencement of and during processing, the inspector should pay attention to the state of raw materials, the preparation of raw materials for processing (peeling, slicing, dicing, blanching, etc.), preparation and density of packing medium (sugar syrup, salt brine, etc.), the state of cans or containers to be used (cleanliness and strength), the cooking or freezing process (time/temperature relationship), can filling and closure and can/container storage.

After processing, the inspector should check the final product to ensure the drained or thawed weight, the vacuum and headspace, packing medium strength and that can/container conditions are satisfactory. Statistically based sampling plans should be adopted for the examination of final product to ensure it meets the requirements of the export regulations. The labelling applied to cans/containers should also be checked to ensure both their correctness and compliance with the export regulations and the requirements of those countries in which the product is to be marketed.

Cans should also be examined to make sure that the correct embossing relating to the product, its date of production and the registered number of the export establishment has been applied. Each establishment registered for the export of processed fruit and vegetables or for canned or frozen foods should have its own quality laboratory sufficiently equipped and staffed to carry out physical, chemical and microbiological examinations of the goods.

Inspectors should have access to the laboratory facilities and the establishment's quality control records as and when required. Independent laboratory examination of product should be made by the agency having responsibility for export on the basis of a statistically developed sampling plan. In those countries where fruit and vegetable production is a seasonal event, processing for export generally takes place at the time of peak production and then declines, often to a halt, as the supply of raw materials declines. As a result, most export establishments produce at their peak of production far more product than they export at that time.

Therefore, most manufacturers find it necessary to store product for considerable periods before it is exported.

Thus, proper storage is essential if the product is to retain its quality and cans remain untarnished. Inspectors should regularly inspect storage facilities, noting their conditions and that of the stored product, looking for signs of deterioration such as pest infestation and rusting of cans.

Prior to export, the exporter should be required to notify the export quality and inspection agency of his intention to export in accordance with the provisions of the export processed fruits and vegetables regulations and on the prescribed "Notice of Intention to Export" form.

The notice should be submitted in sufficient time before the shipment date to enable the product to be inspected satisfactorily; the intensity of inspection depending on the original state of the product, the conditions under which it has been stored and the length of storage. When product is approved, the agency will issue the exporter an "Export Permit" authorizing Custom's clearance of the product.

Export Quality Control and Inspection Systems for Foods

With the advent and development of a food consciousness amongst consumers, stimulated by the work of the Joint FAO/WHO Codex Alimentarius Commission through its elaboration of food standards, codes of hygienic practice and the Code of Ethics for International Trade in Food, an increasing number of countries have adopted sophisticated food laws and established food control agencies, some with the aid of FAO. Consequently, those countries no longer accept products on trust that they are satisfactory, but instead, demand that food imports meet the requirements of their food laws and pass inspection by their control agencies. Moreover, many of them require exporting countries to certify that products comply with their national legislation and some also require additional special declarations.

As a result of these developments the emphasis of activity of Export Quality Control and Inspection Systems has changed. Although most of them still establish their own standards of quality control and adopt standards for foods for export, most of their effort and resources are now directed at ensuring that foods for export meet the mandatory requirements of importing countries and providing the necessary associated certification. To do otherwise is to invite either the detention or, at worst, rejection of product at point of entry.

Detentions and Rejections

Food exporting countries can no longer assume that there is a good chance that products not complying with the requirements of importing countries will escape the inspection at the point of entry.

Details of foods imports released by the United States Food and Drug Administration (FDA) indicate that significant quantities of product are at least detained, and at worst rejected, because they fail to meet U.S. food laws.

Reasons given for the detentions include:

- non compliance with labelling requirements;
- decomposition;

- insect and animal filth and damage;
- use of prohibited additives;
- non compliance with requirements of the U.S. low acid canned food regulations;
- heavy metal contamination;
- excessive levels of pesticide residues;
- excessive levels of mycotoxin;
- mould infestation;
- microbiological contamination;
- swollen and otherwise faulty cans.

The message for food exporting countries is quite clear - ensure your products comply with the mandatory requirements of importing countries or run the very real risk of having them rejected at considerable financial loss to the exporter and the country and resulting in damage to the commercial reputation of both.

While the foregoing relates to the U.S.A. experience, because it is the only country that currently publishes data about detentions and rejections of food imports it can be assumed that record more or less reflects the experience of other food importing countries. It might well be asked why such significantly high levels of detentions and rejections of food imports take place. Undoubtedly the reasons are many and varied. However, the evidence shows that the most important reasons include:

- the inability of some export food industries, especially in developing countries, to handle, process, package and transport products to meet the mandatory requirements of importing countries;
- lack of awareness by food exporting countries of the mandatory requirements of importing countries, including certification;
- lack of adequate export control programmes and related agencies in food exporting countries, preventing them from exercising the necessary product surveillance and giving reliable and credible certification, and
- a lack of communication, between food control authorities and agencies in exporting and importing countries.

All four can be remedied by governments if they possess sufficient political will and take the necessary steps to do so.

Good Manufacturing Practices (GMP); Hygiene Requirements

Any person who has an illness, open lesions, including boils, sores, infected wounds, or any other abnormal source of microbial contamination must not work in any operation (in a food processing centre) which could result in the food, food-contact surface, or food packaging materials becoming contaminated.

Cleanliness

The following applies to people who work in direct contact with food preparation, food ingredients or surfaces of equipment or utensils that will contact food: they must wear clean outer garments, maintain a high degree of personal cleanliness and conform to hygienic practices while on duty; they must wash their hands thoroughly and, if they are working at a job where it is necessary, they must also sanitize their hands before starting work, after each absence from the workstation and at any other time when the hands have become soiled or contaminated; they must also remove all unsecured jewellery. People who are actually handling food, should remove any jewellery that cannot be properly sanitized from their hands; it is necessary to wear effective hair restraints, such as hairnets, caps, headbands or beard covers; operators must not store clothing or other personal belongings in food processing areas. Also, eating food, drinking beverage or using tobacco (in any form) must not be allowed in food processing area; all necessary steps have to be taken by supervisors to prevent operators from contaminating foods with microorganisms or foreign substances such as perspiration, hair, cosmetics, tobacco, chemicals and medicants.

Education and Training

Persons who are monitoring the sanitation programs must have the education and/or experience to demonstrate that they are qualified. Food handlers and supervisors should receive training that will make them aware of the danger of poor personal hygiene and unsanitary work habits.

Supervision

Someone must be assigned the responsibility that all personnel will comply with all the requirements of these GMP's.

Plants and Grounds

Grounds around a food processing centre which are under the control of this centre must be free from conditions such as: improperly stored equipment; litter, waste or refuse; uncut weeds or grass close to buildings; excessively dusty roads, yards or parking lots; inadequately drained areas - potential foot-borne filth or breeding places for insects or microorganisms; inadequately operated systems for waste treatment and disposal.

Plant construction and design shall: provide enough space for sanitary arrangement of equipment and storage of materials; floors, walls and ceilings must be constructed so that they are cleanable and must be kept clean and in good repair; separate by partition, location, time and other means, any operations that may cause cross-contamination of food products with undesirable microorganisms, chemicals, filth or other extraneous material; provide effective screening or other protection to keep out birds, animals and vermin such as insects and rodents.

Provide adequate ventilation to prevent contamination of foods with odours, noxious fumes or vapours (including steam); light bulbs, skylights or any other glass must be of the safety type or protected so that glass contamination cannot occur in case of breakage.

Sanitary Operations

General maintenance. The plant and all fixtures must be kept in good repair and be maintained in a sanitary condition. Cleaning operations must be conducted in a manner that will minimize the possibility of contaminating foods or equipment surfaces that contact food.

Pest control:

- No animals or birds are allowed anywhere in the plant
- Programs must be in effect to prevent contamination by animals, birds and pests, such as rodents and insects;
- Insecticides and rodenticides may be used as long as they are used properly (according to label instructions);
- These pesticides must not contaminate food or packaging materials with illegal residues;

Sanitation of equipment and utensils:

- Utensils and equipment surfaces that are in contact with food must be cleaned as often as necessary to prevent food contamination;
- Equipment surfaces that are not in contact with food should be cleaned as frequently as necessary to minimize accumulation of dust, dirt, food particles, etc.
- Single-service articles such as disposable utensils, paper cups, paper towels, etc., should be:
- Stored in appropriate containers;
- Handled, dispensed, used and disposed of in a manner that prevents contamination of food or equipment;
- Where there is the possibility of introducing undesirable microorganisms into food, all utensils and equipment surfaces that contact food must be cleaned and sanitized before use and following any interruption during which they may have become contaminated;

- When utensils or equipment are used in a continuous production operation, they must be cleaned and sanitized on a predetermined schedule;
- Any facility, procedure, machine or device may be used for cleaning and sanitizing, as long as it has been established that the procedure will do the job effectively.

Storage and handling of clean portable equipment and utensils:

a) This refers to portable equipment or utensils which have surfaces that will contact foods;

b) When such equipment or utensils have been cleaned and sanitized, they should be stored in a manner that will protect the food contact surfaces from splash, dust and other contamination.

Sanitary Facilities and Controls

Water supply. Any water that comes into contact food or processing equipment must be safe and of adequate sanitary quality.

Sewage disposal. Must flow into an adequate sewage system or disposed of through other adequate means.

Plumbing. Must be of adequate size and design to:

a. Supply enough water to areas in the plant where it is needed;
b. Properly convey sewage or disposable liquid waste from the plant;
c. Not create a source of contamination or unsanitary condition;
d. Provide adequate floor drainage where hosing-type cleaning is done or where operations discharge water or liquid waste onto the floor;
e. Insure that there is no backflow from cross-connection between piping systems that discharge waste water or sewage, and those that carry water for food or food manufacturing.

Toilet facilities:

a. Toilets and hand-washing facilities must be provided inside the fruit and vegetable processing centres;
b. Toilet tissue must be provided;
c. Toilets must be kept sanitary and in good repair;
d. Toilet rooms must have self-closing doors;
e. Toilet rooms must not open directly into areas where food is exposed unless steps have been taken to prevent airborne contamination (example: double doors, positive airflow, etc.);
f. Signs must be posted that direct employees to wash their hands with soap or detergent after using the toilet.

Hand-washing Facilities

Adequate and convenient hand-washing and, if necessary, hand-sanitizing facilities must be provided anywhere in the plant where the nature of employees jobs requires that they wash, sanitize and dry their hands;

These hand-washing facilities must provide:

- Running water at a suitable temperature;
- Effective hand-cleaning and hand-sanitizing preparations;
- Clean towel service or suitable drying devices;
- Easily cleanable waste receptacle;
- Water control valves designed and constructed to protect against recontamination of clean, sanitized hands;
- Signs directing employees handling unprotected food to wash and, if appropriate, sanitize theirs hands before starting work, after each absence from the workstation, and any other time when the hands have become soiled or contaminated.

Rubbish and offal disposal must be handled in such a manner that they do not serve to attract or harbour pests or create contaminating conditions.

Equipment and utensils:

a. Equipment and utensils must be designed and constructed so that they are adequately cleanable and will not adulterate food with lubricants, fuel, metal fragments, contaminated water, etc.

b. Equipment should be installed so that it, and the area around it, can be cleaned;

c. Food contact surfaces shall be made of nontoxic materials and must be corrosion-resistant;

d. Seams on food contact surfaces shall be smoothly bonded, or maintained in order to minimize the accumulation of food particles, dirt and organic matter;

e. Equipment in processing areas that does not come into contact with food shall be constructed so that it can be kept clean;

f. Holding, conveying and manufacturing systems, including gravimetric, pneumatic, closed and automated systems, shall be maintained in a sanitary condition;

g. Each freezer and cold storage compartment shall have an indicating thermometer, temperature measuring or recording device, and should have an automatic control for regulating temperature, or an automatic alarm system to indicate a significant temperature change;

h. Instruments and controls used for measuring, regulating or recording temperatures, pH, acidity, water activity, etc. shall be adequate in number, accurate and maintained.

Processes and Controls

There must be an individual who is responsible for supervising the overall sanitation of the plant.

Raw materials and ingredients:

a. Must be inspected and sorted to insure that they are clean, wholesome and fit for processing into human food;

b. Must be stored under conditions that will protect against contamination and minimize deterioration;

c. Must be washed or cleaned to remove soil and other contamination:

   - Water used for washing, rising or conveying food products must be of sanitary quality;
   - Water must not be reused for washing, rinsing or conveying if contamination of food may result;
   - Containers and carriers (such as trucks or railcars) should be inspected to assure that their condition has not contaminated raw ingredients;

d. Raw materials shall not contain levels of microorganisms that may produce food poisoning or other disease, or they shall be pasteurized or otherwise treated during manufacturing operations so that the product will not be adulterated;

e. Materials susceptible to contamination with natural toxins, e.g., aflatoxin, shall comply with national and international official levels before they are incorporated into the finished food;

f. Materials susceptible to contamination with pests, undesirable microorganisms, or extraneous material, shall comply with national and international regulations, guidelines and defect action levels;

g. Materials shall be stored in containers, and under conditions which protect against contamination;

h. Frozen materials shall be kept frozen. If thawing is required prior to use, it shall be done in a manner that prevents contamination.

Manufacturing Operations

a. Food processing equipment must be keep in a sanitary condition through frequent cleaning and, when necessary, sanitizing. If necessary, such equipment must be taken apart for thorough cleaning.

b. It is necessary to process, package and store food under conditions that will minimize the potential for undesirable microbiological growth, toxin formation, deterioration or contamination. To accomplish this may require careful monitoring of such factors as time, temperature, humidity, pressure, flow rate, etc. The object is to assure that mechanical breakdowns, time delays, temperature fluctuations or other factors do not allow the foods to decompose or become contaminated.

c. Food shall be held under conditions that prevent the growth of undesirable microorganisms as follows:
   - Refrigerated foods shall be maintained at 45° F or below;
   - Frozen foods shall be maintained in a frozen state;
   - Acid or acidified foods to be held in hermetically sealed containers at ambient temperatures shall be heat-treated to destroy mesophyllic microorganisms;

d. Measures such as sterilizing, irradiating, pasteurizing, etc., shall be adequate to destroy or prevent the growth of undesirable microorganisms;

e. Work-in-process shall be protected against contamination;

f. Finished food shall be protected from contamination;

g. Equipment, containers and utensils shall be constructed, handled and maintained to protect against contamination;

h. Measures, e.g., sieves, traps, metal detectors, shall be used to protect against the inclusion of metal or other extraneous material in food;

i. Food or materials that are adulterated shall be disposed of in a manner that prevents other food from being contaminated;

j. Mechanical manufacturing steps such as washing, peeling, etc., shall be performed to protect against contamination by providing adequate protection from contaminants that may drip, drain or be drawn into the food, by adequately cleaning and sanitizing all food-contact surfaces and by using time and temperature controls at and between each manufacturing step;

k. Heat-blanching should be done by heating the food to the required temperature, holding it at this temperature for the required time, and then either rapidly cooling the food or passing it to the next manufacturing step without delay;

l. Filling, assembling, packaging, and other operations shall be performed in such a way that the food is protected against contamination by:
   - Use of a quality control operation in which the Critical Control Points are identified and controlled during manufacturing;

- Adequate cleaning and sanitizing of all food-contact surfaces and food containers;
- Using materials for food containers and food-packaging materials that are safe and suitable;
- Providing physical protection from contamination, particularly airborne contamination;
- Using sanitary handling procedures.

m. Food such as, but not limited to, dry mixes, nuts, intermediate moisture food, and dehydrated food, that relies on the control of aw for preventing the growth of undesirable microorganisms shall be processed to and maintained at a safe moisture level by:

- Monitoring the aw of food;
- Controlling the soluble solids/water ratio in finished food;
- Protecting finished food from moisture pick-up, by use of a moisture barrier, or by other means, so that the Aw of the food does not increase to an unsafe level;

n. Food such as, but not limited to, acid and acidified food, that relies principally on the control of pH for preventing the growth of undesirable microorganisms shall be monitored and maintained at a pH of 4.6 or below by: - Monitoring the pH of raw materials, food in process, and finished food; - Controlling the amount of acid or acidified food added to low-acid food;

o. If ice is used and comes in contact with food products, it must be made from potable water and be in a sanitary condition;

p. Areas and equipment that are used to process human food should not be used to process non-human food-grade animal feed, or inedible products unless there is no possibility of contaminating the human food;

q. A coding system should be utilized that will allow positive lot identification in the event it is necessary to identify and segregate lots of food that may be contaminated.

- Records should be kept for a period of time that exceeds the self life of the product, except that
- Records need not be kept beyond two years.

8

Milk History, Consumption, Production, and Composition

This course is about the study of milk and milk-derived food products from a food science perspective. It focuses on the biological, chemical, physical, and microbiological aspects of milk itself, and on the technological (processing) aspects of the transformation of milk into its various consumer products, including beverages, fermented products, concentrated and dried products, butter and ice cream.

Milk is as ancient as mankind itself, as it is the substance created to feed the mammalian infant. All species of mammals, from man to whales, produce milk for this purpose. Many centuries ago, perhaps as early as 6000-8000 BC, ancient man learned to domesticate species of animals for the provision of milk to be consumed by them. These included cows (genus *Bos*), buffaloes, sheep, goats, and camels, all of which are still used in various parts of the world for the production of milk for human consumption.

Fermented products such as cheeses were discovered by accident, but their history has also been documented for many centuries, as has the production of concentrated milks, butter, and even ice cream.

Technological advances have only come about very recently in the history of milk consumption, and our generations will be the ones credited for having turned milk processing from an art to a science. The availability and distribution of milk and milk products today in the modern world is a blend of the centuries old knowledge of traditional milk products with the application of modern science and technology.

The role of milk in the traditional diet has varied greatly in different regions of the world. The tropical countries have not been traditional milk consumers, whereas the more northern regions of the world, Europe (especially Scandinavia) and North America, have traditionally consumed far more milk and milk products in their diet. In tropical countries where high temperatures and lack of refrigeration has led to the inability to produce and store fresh milk, milk has traditionally been preserved through means other than refrigeration, including immediate consumption of warm milk after milking, by boiling milk, or by conversion into more stable products such as fermented milks.

World-wide Milk Consumption and Production

Table: *Per Capita Consumption of Milk and Milk Products in Various Countries, 2006 data.*

Country	*Liquid Milk Drinks (Litres)*	*Cheeses (kg)*	*Butter (kg)*
Finland	183.9	19.1	5.3
Sweden	145.5	18.5	1.0
Ireland	129.8	10.5	2.9
Netherlands	122.9	20.4	3.3
Norway	116.7	16.0	4.3
Spain (2005)	119.1	9.6	1.0
Switzerland	112.5	22.2	5.6
United Kingdom (2005)	111.2	12.2	3.7
Australia (2005)	106.3	11.7	3.7
Canada (2005)	94.7	12.2	3.3
European Union (25 countries)	92.6	18.4	4.2
Germany	92.3	22.4	6.4
France	92.2	23.9	7.3
New Zealand (2005)	90.0	7.1	6.3
United States	83.9	16.0	2.1
Austria	80.2	18.8	4.3
Greece	69.0	28.9	0.7
Argentina (2005)	65.8	10.7	0.7
Italy	57.3	23.7	2.8
Mexico	40.7	2.1	N/A
China (2005)	8.8	N/A	N/A

Source: International Dairy Federation, Bulletin 423/2007.

The total milk consumption (as fluid milk and processed products) per person varies widely from highs in Europe and North America to lows in Asia. However, as the various regions of the world become more integrated through travel and migration, these trends are changing, a factor which needs to be considered by product developers and marketers of milk and milk products in various countries of the world.

Table: *Cow milk production ('000 tonnes) in selected countries in the world (2006).*

United States	82,462
India	39,759
China	31,934
Russia	31,100
Germany	27,955
Brazil	25,750
France	24,195
New Zealand	15,000
United Kingdom	14,359
Ukraine	13,287
Poland	11,970
Italy	11,186
Netherlands	10,995
Mexico	10,352
Argentina	10,250
Turkey	10,000
Australia	9550
Canada	7854

Source: International Dairy Federation, Bulletin 423/2007.

Even within regions such as Europe, the custom of milk consumption has varied greatly. Consider for example the high consumption of fluid milk in countries like Finland, Norway and Sweden compared to France and Italy where cheeses have tended to dominate milk consumption. When you also consider the climates of these regions, it would appear that the culture of producing more stable products (cheese) in hotter climates as a means of preservation is evident. Table illustrates milk per capita consumption information from various countries of the world. Table shows the quantity of raw milk produced around the world.

Milk Composition

The role of milk in nature is to nourish and provide immunological protection for the mammalian young. Milk and honey are the only articles of diet whose sole function in nature is food. It is not surprising, therefore, that the nutritional value of milk is high.

Table: *Composition of Milk from Different Mammalian Species (per 100 g fresh milk).*

	Protein (g)	Fat (g)	Carbohydrate (g)	Energy (kcal)
Cow	3.2	3.7	4.6	66
Human	1.1	4.2	7.0	72
Water Buffalo	4.1	9.0	4.8	118
Goat	2.9	3.8	4.7	67
Donkey	1.9	0.6	6.1	38
Elephant	4.0	5.0	5.3	85
Monkey, rhesus	1.6	4.0	7.0	73
Mouse	9.0	13.1	3.0	171
Whale	10.9	42.3	1.3	443
Seal	10.2	49.4	0.1	502

Source: Webb, B.H., A.H. Johnson and J.A. Alford. 1974. Fundamentals of Dairy Chemistry. Second Ed. AVI Publishing Co., Westport, CT., Chap. 1.

Table: *Gross composition of milk of various breeds, g/100g.*

	Body Wt.	Milk Yield	Fat	Protein	Lactose	Ash	Total Solids
	(kg)	(kg)	(%)	(%)	(%)	(%)	(%)
Holstein	640	7360	3.54	3.29	4.68	0.72	12.16
Brown Swiss	640	6100	3.99	3.64	4.94	0.74	13.08
Ayrshire	520	5760	3.95	3.48	4.60	0.72	12.77
Guernsey	500	5270	4.72	3.75	4.71	0.76	14.04
Jersey	430	5060	5.13	3.98	4.83	0.77	14.42
Shorthorn	530	5370	4.00	3.32	4.89	0.73	12.9

Holstein: 12.16% T.S. x 7360 kg/lactation = 895 kg of total solids produced/ lactation (140% of her body wt.!)

Jersey: 14.42% T.S. x 5060 kg/lactation = 730 kg of total solids produced/ lactation (170% of her body wt.!)

Source: Webb, B.H., A.H. Johnson and J.A. Alford. 1974. Fundamentals of Dairy Chemistry. Second Ed. AVI Publishing Co., Westport, CT., Chap. 1.

Milk Production and Biosynthesis

Milk Production

Milk is the source of nutrients and immunological protection for the young cow. The gestation period for the female cow is 9 months. Shortly before calving, milk is secreted into the udder in preparation for the new born. At parturition, fluid from the mammary gland known as colostrum is secreted. This yellowish coloured, salty liquid has a very high serum protein content and provides antibodies to help protect the newborn until its own immune system is established. Within 72 hours, the composition of colostrum returns to that of fresh milk, allowing to be used in the food supply.

The period of lactation, or milk production, then continues for an average of 305 days, producing as much as 9000 or more kg of milk. This is quite a large amount considering the calf only needs about 1000 kg for growth.

Within the lactation, the highest yield is 2-3 months post-parturition, yielding 40-50 L/day. Within the milking lifetime, a cow reaches a peak in production about her third lactation, but can be kept in production for 5-6 lactations if the yield is still good.

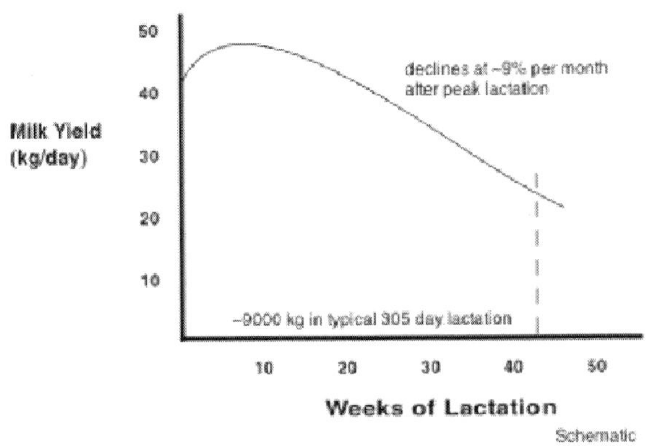

About 1-2 months after calving, the cow begins to come into heat again. She is usually inseminated about 3 months after calving so as to come into a yearly calving cycle. Heifers are normally first inseminated at 15 months so she's 2 when the first calf is born. About 60 days before the next calving, the cow is dried off. There is no milking during this stage for two reasons:

1. milk has tapered off because of maternal needs of the fetus
2. udder needs time to prepare for the next milking cycle

The life of a female cow can be summerized as follows:

Age

0	*Calf born*
15 mos	Heifer inseminated for first calf
24 mos	First calf born-starts milking
27 mos	Inseminated for second calf
34 mos	Dried off
36 mos	Second calf born-starts milking

Automatic Milking

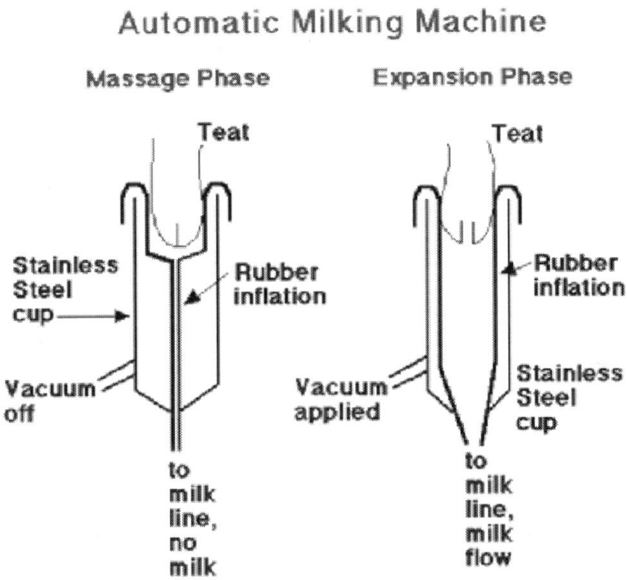

Effects of Milk Handling on Quality and Hygiene

Cleanliness

The environment of production has a great effect on the quality of milk produced. From the food science perspective, the production of the highest quality milk should be the goal. However, this is sometimes not the greatest concern of those involved in milk production. Hygienic quality assessment tests include sensory tests, dye reduction tests for microbial activity, total bacterial count (standard plate count), sediment, titratable acidity, somatic cell count, antibiotic residues, and added water. The two common dye reduction tests are methylene blue and resazurin. These are both synthetic compounds which accept electrons and change colour as a result of this reduction. As part of natural metabolism, active microorganisms transfer electrons, and thus rate at which dyes added to milk are reduced is an indication of the level of microbial activity. Methylene blue turns from blue to colorless, while resazurin turns from blue to violet to pink to colourless. The reduction time is inversely correlated to bacterial numbers. However, different species react differently. Mesophilics are favoured over psychrotrophs, but psychrotrophic organisms tend to be more numerous and active in cooled milk.

Temperature

Milk production and distribution in the tropical regions of the world is more challenging due to the requirements for low-temperature for milk stability. Consider the following chart illustrating the numbers of bacteria per millilitre of milk after 24 hours:

5°C	2,600
10°C	11,600
12.7°C	18,800
15.5°C	180,000
20°C	450,000

Traditionally, this has been overcome in tropical countries by stabilizing milk through means other than refrigeration, including immediate consumption of warm milk after milking, by boiling milk, or by conversion into more stable products such as fermented milks.

Mastitis and Antibiotics

Mastitis is a bacterial and yeast infection of the udder. Milk from mastitic cows is termed abnormal. Its SNF, especially lactose, content is decreased, while Na and Cl levels are increased, often giving mastitic milk a salty flavour. The presence of mastitis is also accompanied by increases in bacterial numbers, including the possibility of human pathogens, and by a dramatic increase in somatic cells.

These are comprised of leukocytes (white blood cells) and epithelial cells from the udder lining. Increased somatic cell counts are therefore indicative of the presence of mastitis. Once the infection reaches the level known as "clinical' mastitis, pus can be observed in the teat canal just prior to milking, but at sub-clinical levels, the presence of mastitis is not obvious.

Antibiotics are frequently used to control mastitis in dairy cattle. However, the presence of antibiotic residues in milk is very problematic, for at least three reasons. In the production of fermented milks, antibiotic residues can slow or destroy the growth of the fermentation bacteria.

Somatic Cell Count (000's/ml)	***Daily Milk Yield (kg)***	
	1st Lactation	***Older Lactations***
0-17	23.1	29.3
18-34	23.0	28.7
35-70	22.6	28.0
71-140	22.4	27.4
141-282	22.1	27.0
282-565	21.9	26.3
566-1130	21.4	25.4
1131-2262	20.7	24.6
2263-4525	20.0	23.6
>4526	19.0	22.5

From a human health point of view, some people are allergic to specific antibiotics, and their presence in food consumed can have severe consequences. Also, frequent exposure to low level antibiotics can cause microorganisms to become resistant to them, through mutation, so that they are ineffective when needed to fight a human infection. For these reasons, it is extremely important that milk from cows being treated with antibiotics is withheld from the milk supply. The withdrawal time after final treatment for various antibiotics is shown below:

- Amoxcillin 60 hrs.
- Cloxacillin 48 hrs.
- Erythromicin 36 hrs.
- Novobiocin 72 hrs.
- Penicillin 84 hrs.
- Sulfadimethozine 60 hrs.
- Sulfabromomethozine 96 hrs.
- Sulfaethoxypyridozine 72 hrs.

Anti-Microbial Systems in Raw Milk

There exists in milk a number of natural anti-microbial defence mechanisms. These include:

- lysozyme-an enzyme that hydrolyses glycosidic bonds in gram positive cell walls. However, its effect as a bacteriostatic mechanism in milk is probably negligible.
- lactoferrin-an iron binding protein that sequesters iron from microorganisms, thus taking away one of their growth factors. Its effect as a bacteriostatic mechanism in milk is also probably negligible.

- lactoperoxidase-an enzyme naturally present in raw milk that catalyzes the conversion of hydrogen peroxide to water. When hydrogen peroxide and thiocyanate are added to raw milk, the thiocyanate is oxidized by the enzyme/ hydrogen peroxide complex producing bacteriostatic compounds that inhibit Gram negative bacteria, E. coli, Salmonella spp, and streptococci. This technique is being used in many parts of the world, especially where refrigeration for raw milk is not readily available, as a means of increasing the shelf life of raw milk.

Milk Biosynthesis

Milk is synthesized in the mammary gland. Within the mammary gland is the milk producing unit, the alveolus. It contains a single layer of epithelial secretory cells surrounding a central storage area called the lumen, which is connected to a duct system. The secretory cells are, in turn, surrounded by a layer of myoepithelial cells and blood capillaries.

The raw materials for milk production are transported via the bloodstream to the secretory cells. It takes 400-800 L of blood to deliver components for 1 L of milk.

- Proteins: building blocks are amino acids in the blood. Casein micelles, or small aggregates thereof, may begin aggregation in Golgi vesicles within the secretory cell.
- Lipids:
  - o C4-C14 fatty acids are synthesized in the cells
  - o C16 and greater fatty acids are preformed as a result of rumen hydrogenation and are transported directly in the blood
- Lactose: milk is in osmotic equilibrium with the blood and is controlled by lactose, K, Na, Cl; lactose synthesis regulates the volume of milk secreted

The milk components are synthesized within the cells, mainly by the endoplasmic reticulum (ER) and its attached ribosomes. The energy for the ER is supplied by the mitochondria.

The components are then passed along to the Golgi apparatus, which is responsible for their eventual movement out of the cell in the form of vesicles. Both vesicles containing aqueous non-fat components, as well as liquid droplets (synthesized by the ER) must pass through the cytoplasm and the apical plasma membrane to be deposited in the lumen. It is thought that the milk fat globule membrane is comprised of the apical plasma membrane of the secretory cell. Milking stimuli, such as a sucking calf, a warm wash cloth, the regime of parlour etc., causes the release of a hormone called oxytocin. Oxytocin is relased from the pituitary gland, below the brain, to begin the process of milk let-down.

As a result of this hormone stimulation, the muscles begin to compress the alveoli, causing a pressure in the udder known as letdown reflex, and the milk components stored in the lumen are released into the duct system. The milk is forced down into the teat cistern from which it is milked. The let-down reflex fades as the oxytocin is degraded, within 4-7 minutes. It is very difficult to milk after this time.

Milk Grading and Defects

The importance of milk grading lies in the fact that dairy products are only as good as the raw materials from which they were made. It is important that dairy personnel have a knowledge of sensory perception and evaluation techniques. The identification of off-flavours and desirable flavours, as well as knowledge of their likely cause, should enable the production of high quality milk, and subsequently, high quality dairy products.

Milk Grading

- Sense of Taste
- Sense of Smell
- Techniques.

Milk Defects

- Lipolyzed
- Oxidiation
- Sunlight
- Cooked
- Transmitted
- Microbial.

Milk Grading

An understanding of the principles of sensory evaluation are neccessary for grading. All five primary senses are used in the sensory evaluation of dairy products: sight, taste, smell, touch and sound. The greatest emphasis, however, is placed on taste and smell.

The Sense of Taste

Taste buds, or receptors, are chiefly on the upper surface of the tongue, but may also be present in the cheek and soft palates of young people. These buds, about 900 in number, must make contact with the flavouring agent before a taste sensation occurs. Saliva, of course, is essential in aiding this contact. There are four different types of nerve endings on the tongue which detect the four basic "mouth" flavours-sweet, salt, sour, and bitter. Samples must, therefore, be spread around in the mouth in order to make positive flavour identification. In addition to these basic tastes, the mouth also allows us to get such reactions as coolness, warmth, sweetness, astringency, etc.

The Sense of Smell

We are much more perceptive to the sense of smell than we are to taste. For instance, it is possible for an odouriferous material such as mercaptain to be detected in 20 billion parts of air. The centres of olfaction are located chiefly in the uppermost part of the nasal cavity. To be detectable by smell, a substance must dissolve at body temperature and be soluble in fat solvents.

Note: The sense of both taste and smell may become fatigued during steady use. A good judge does not try to examine more than one sample per minute. Rinsing the mouth with water between samples may help to restore sensitivity.

Milk Grading Techniques

Temperature should be between 60-70° F (15.5-21° C) so that any odour present may be detected readily by sniffing the container. Also, we want a temperature rise when taking the sample into the mouth; this serves to volatize any notable constituents. Noting the odour by placing the nose directly over the container immediately after shaking and taking a full "whiff" of air. Any off odour present may be noted. Need to make sure we have a representative sample; mixing and agitation are important. Agitation leaves a thin film of milk on the inner surface which tends to evaporate giving off odour if present.

During sampling, take a generous sip, roll about the mouth, note flavour sensation, and expectorate. Swallowing milk is a poor practice. Can enhance the after-taste by drawing a breath of fresh air slowly through the mouth and then exhale slowly through the nose. With this practice, even faint odours can be noted.

Milk has a flavour defect if it has an odour, a foretaste or an aftertaste, or does not leave the mouth in a clean, sweet, pleasant condition after tasting.

Characterization of Flavour Defects-ADSA

Lipolytic or Hydrolytic Rancidity

Rancidity arises from the hydrolysis of milkfat by an enzyme called the lipoprotein lipase (LPL). The flavour is due to the short chain fatty acids produced, particularly butyric acid. LPL can be indigenous or bacterial. It is active at the fat/water interface but is ineffective unless the fat globule membrane is damaged or weakened. This may occur through agitation, and/or foaming, and pumping. For this reason, homogenized milk is subject to rapid lipolysis unless lipase is destroyed by heating first; the enzyme (protein) is denatured at 55-60° C. Therefore, always homogenize milk immediately before or after pasteurization and avoid mixing new and homogenized milk because it leads to rapid rancidity.

Some cows can produce spontaneous lipolysis from reacting to something indigenous to the milk. Late lactation, mastitis, hay and grain ratio diets (more so than fresh forage or silage), and low yielding cows are more suseptible. Lipolysis can be detected by measuring the acid degree value which determines the presence of free fatty acids. Lipolytic or hydrolytic rancidity is distinct from oxidative rancidity, but frequently in other fat industries, rancid is used to mean oxidative rancidity; in dairy, rancidity means lipolysis. Characterized: soapy, blue-cheese like aroma, slightly bitter, foul, pronounced aftertaste, does not clear up readily

Oxidation

Milk fat oxidation is catalysed by copper and certain other metals with oxygen and air. This leads to an autooxidation reaction consisting of initiation, propagation, termination.

RH — R + H initiation-free radical

$R + O_2$ —— RO_2 propagation

RO_2 + RH — ROOH + R

R + R — R_2 termination

R + RO_2 — RO_2R

It is usually initiated in the phospholipid of the fat globule membrane. Propagation then occurs in triglycerides, primarily double bonds of unsaturated fatty acids. During propagation, peroxide derivatives of fatty acids accumulate. These undergo further reactions to form carbonyls, of which some, like aldehydes and ketones, have strong flavours. Dry feed, late lactation, added copper or other metals, lack of vit E (tocopherol) or selenium (natural antioxidates) in the diet all lead to spontaneous oxidation. It can be a real problem especially in winter. Exposure to metals during processing can also contribute.

Characterized: metallic, wet cardboard, oily, tallowy, chalky; mouth usually perceives a puckery or astringent feel

Sunlight

Often confused with oxidized, this defect is caused by UV-rays from sunlight or flourescent lighting catalyzing oxidation in unprotected milk. Photo-oxidation activates riboflavin which is responsible for catalyzing the conversion of methionine to methanal. It is, therefore, a protein reaction rather than a lipid reaction. However, the end product flavour notes are similar but tends to diminish after storage of several days.

Characterized: burnt-protein or burnt-feathers-like, "medicinal"-like flavour.

Cooked

This defect is a function of the time-temperature of heating and especially the presence of any "burn-on" action of heat on certain proteins, particulary whey proteins. Whey proteins are a source of sulfide bonds which form sulfhydryl groups that contribute to the flavour. The defect is most obvious immediately after heating but dissipates within 1 or 2 days.

Characterized: slightly cooked or nutty-like to scorched or caramelized.

Transmitted flavours

Cows are particulary bad for transmitting flavours through milk and milk is equally as susceptible to pick-up of off flavours in storage. Feed flavours and green grass can be problems so it is necessary to remove cows from feed 2-4 hrs before milking. Weeds, garlic/onion, and dandelions can tranfer flavours to the milk and even subsequent products such as butter. Barny flavours can be picked up in the milk if there is poor ventilation and the barn is not properly cleared and cows breathe the air. These flavours are volatile so can be driven off through vacuum de-aeration.

Characterization: hay/silage, cowy/barny.

Microbial

There are many flavour defects of dairy products that may be caused by bacteria, yeasts, or moulds. In raw milk the high acid/sour flavour is caused by the growth of lactic acid bacteria which ferment lactose. It is less common today due to change in raw milk microflora. In both raw or processed milk, fruity flavours may arise due to psychrotrophs such as *Pseudomonas fragi*. Bitter or putrid flavours are caused by psychrotrophic bacteria which produce protease. It is the proteolytic action of protease that usually causes spoilage in milk. Malty flavours are caused by *S.lactis* var. *maltigenes* and is characterized by a corn flakes type flavour. Although more of a tactile defect, ropy milk is also caused by bacteria, specifically those which produce exopolysaccharides.

Miscellaneous Defects

- astringent
- chalky
- chemical/medicinal-disease-associated or adulteration

- flat-adulteration (water)
- foreign
- salty-disease associated
- bitter-adulteration.

Milk flavour is graded on a score of one to 10. Some flavour defects, even if only slightly present, can decrease the score drastically. The following are suggested flavour scores for milk with designated intensities of flavour defects.

Flavour Criticisms	***Intensity of Defect***		
	Slight	***Definite***	***Pronounced***
Astringent	8	7	5
Barny	7	5	3
Bitter	7	5	3
Cooked	9	8	6
Cowy	6	4	1
Feed	9	7	5
Flat	9	8	7
Foreign	5	3	0
Garlic/onion	5	3	1
High acid	3	1	0
Bacterial	5	3	0
Lacks Freshness	7	5	3
Malty	7	5	3
Oxidized	7	5	3
Rancid	7	5	3
Salty	8	6	4
Unclean	7	5	3

9

Condensed Milk and Milk Powder

Condensed Milk

Condensed milk, also known as sweetened condensed milk, is cow's milk from which water has been removed and to which sugar has been added, yielding a very thick, sweet product that can last for years without refrigeration if unopened. The two terms, condensed milk and sweetened condensed milk, have become synonymous; though there have been unsweetened condensed milk products, today these are uncommon.

History

According to the writings of Marco Polo, the Tartars were able to condense milk. Ten pounds (4.54 kg) of milk paste was carried by each man who would mix the product with water. However, this probably refers to the soft Tartar curd which can be made into a drink ("airan") by diluting it and therefore to fermented, not fresh, milk concentrate.

Nicolas Appert condensed milk in France in 1820. and was developed after in the United States in 1856 by Gail Borden, Jr. in reaction to difficulties in storing milk for more than a few hours. Before this development milk could only be kept fresh for a short while and was only available in the immediate vicinity of a cow. While returning from a trip to England in 1851, Borden was devastated by the death of several children, apparently due to poor milk from shipboard cows. With less than a year of schooling and following in a wake of failures, both of his own and of others, Borden was inspired by the vacuum pan he had seen being used by Shakers to condense fruit juice and was at last able to reduce milk without scorching or curdling it. Even then his first two factories failed and only the third, built with his new partner, Jeremiah Milbank in Wassaic, New York, produced a usable milk derivative that was long-lasting and needed no refrigeration.

Probably of equal importance for the future of milk were Borden's requirements (the "Dairyman's Ten Commandments") for farmers who wanted to sell him raw milk: they were required to wash udders before milking, keep barns swept clean, and scald and dry their strainers morning and night. By 1858 Borden's milk, sold as Eagle Brand, had gained a reputation for purity, durability and economy.

In 1864, Gail Borden's New York Condensed Milk Company constructed the New York Milk Condensery in Brewster, New York. This condensery was the largest and most advanced milk factory and was Borden's first commercially successful plant. Over 200 dairy farmers supplied 20,000 gallons (76,000 litres) of milk daily to the Brewster plant as demand was driven by the Civil War.

The U.S. government ordered huge amounts of it as a field ration for Union soldiers during the American Civil War. This was an extraordinary field ration for the nineteenth century: a typical 14 oz (400 g) can contains 1,300 calories (5440 kj), 1 oz (30 g) each of protein and fat, and more than 7 oz (200 g) of carbohydrate.

Soldiers returning home from the Civil War soon spread the word. By the late 1860s, condensed milk was a major product. The first Canadian condensery was built at Truro, Nova Scotia, in 1871. In 1899, E. B. Stuart opened the first Pacific Coast Condensed Milk Company (later known as the Carnation Milk Products Company) plant in Kent, Washington. Unfortunately, the condensed milk market developed a bubble. Too many manufacturers chased too little demand. By 1912, stocks of condensed milk were large and the price dropped. Many condenseries went out of business. In 1911, Nestlé constructed the world's largest condensed milk plant in Dennington, Victoria, Australia.

In 1914, Professor Otto F Hunziker, head of Purdue University's dairy department, self-published *Condensed milk and milk powder: prepared for the use of milk condenseries, dairy students and pure food departments*. This text, along with additional work of Professor Hunziker and others involved with the American Dairy Science Association, standardized and improved condensery operations in the U.S. and internationally. Hunziker's book was republished in a seventh edition in October 2007 by Cartwright Press.

The first World War regenerated interest in, and a market for, condensed milk, primarily due to its storage and transportation benefits. In the U.S., the higher price for raw milk paid by condenseries created significant problems for the cheese industry.

Production

Raw milk is clarified and standardized, and then is heated to 85-90°C for several seconds. This heating destroys some microorganisms, decreases fat separation and inhibits oxidation. Some water is evaporated from the milk and sugar is added to approximately 45%.

This sugar is what extends the shelf life of sweetened condensed milk. Sucrose increases the liquid's osmotic pressure, which prevents microorganism growth. The sweetened evaporated milk is cooled and lactose crystallization is induced.

Current Use

Condensed milk is used in recipes for the popular Brazilian candy brigadeiro in which condensed milk is the main ingredient (the most famous condensed milk brand in Brazil is Moça , local version of Swiss *Milch Mädchen* marketed by Nestlé), lemon meringue pie, key lime pie, caramel candies and other desserts.

In parts of Asia and Europe, sweetened condensed milk is the preferred milk to be added to coffee or sweetened tea. Many countries in South East Asia use condensed milk to flavour their coffee. A popular treat in Asia is to put condensed milk on toast and eat it in a similar way as jam and toast. Nestlé has even produced a squeeze bottle similar to Smucker's jam squeeze bottles for this very purpose. Condensed milk is a major ingredient in many Indian desserts and sweets. While most Indians start with normal milk to reduce and sweeten it, packaged condensed milk has also become popular.

In New Orleans, it is commonly used as a topping on top of a chocolate or similar cream flavour snowball. In Scotland, it is mixed with sugar and some butter and baked to form a popular, sweet candy called Tablet (confectionery) or Swiss-Milk-Tablet.

In some parts of the Southern U.S., condensed milk is a key ingredient in lemon icebox pie, a sort of cream pie. In the Philippines, condensed milk is mixed with some evaporated milk and eggs, spooned into shallow metal containers over liquid caramelised sugar, then steamed to make a stiffer and more filling version of crème brulée known as leche flan.

During the communism era in Poland it was common to boil a can of condensed milk in water for about 2 hours. The resulting product is called kaymak-sweet semiliquid substance which can be used as a cake icing or put between dry wafers. It is less common nowadays but recently some manufactures of condensed milk introduced canned ready-made kaymak. Boiling the can in this way is central to the making of Banoffee pie and home-made dulce de leche.

Substitutions

To gain condensed milk from 1 cup (250 ml) of evaporated milk one has to add 1 1/4 cups (250 g) of sugar and dissolve it by heating the milk.

Condensed Milk Companies

- Arla Foods (Canned Cream & Milk Company)
- Borden Food Corporation (New York Condensed Milk Company)
- Carnation Evaporated Milk Company (Pacific Coast Condensed Milk Company, Mohawk Milk Company)
- Dairymen's League Cooperative Association
- Foremost Condensery (Western Condensing)
- Milk Producers Cooperative Marketing Company
- Nestlé (Anglo/Swiss Condensed Milk Company)
- Northern Condensed Milk Company
- Northern Foods plc
- Pet Milk Company (Helvetia Milk Condensing Company, Utah Condensed Milk Company)
- Wisconsin Condensed Milk Company (R. G. Fraser & Company).

Baked Milk

Baked milk is a variety of boiled milk that has been particularly popular in Russia and Ukraine. It is made by cooking boiled milk on low heat for eight hours or more.

In rural areas, baked milk has been produced by leaving a jug of boiled milk in an oven for a day or for a night until it is coated with a brown crust. The stove in a traditional Russian loghouse (izba) was designed so as to "sustain varying cooking temperatures based on the placement of the food inside the oven".

Nowadays baked milk is produced on an industrial scale, as is soured or fermented baked milk, traditionally known as ryazhenka. Like scalded milk it is free of bacteria and enzymes, and so can be stored safely at room temperature for up to forty hours. Home-made baked milk was (and still is) used for preparing a range of cakes, pies, and cookies. Caramelized condensed milk is a similar preparation used in homemade pastries. It is sometimes prepared by prolonged heating of the unopened cans with sweetened condensed milk.

Evaporated Milk

Evaporated milk, also known as dehydrated milk, is a shelf-stable canned milk product with about 60% of the water removed from fresh milk. It differs from sweetened condensed milk, which contains added sugar. Sweetened condensed milk requires less processing since the added sugar inhibits bacterial growth. The actual liquid portion of the product takes up half the space of fresh milk. When the non-liquid product is mixed with a proportionate amount of water, evaporated milk becomes the equivalent of fresh milk. This makes evaporated milk attractive for shipping purposes and can have a shelf life of months or even years, depending upon the brand. This made evaporated milk very popular before refrigeration as a safe and reliable substitute for perishable fresh milk, that could be shipped easily to locations lacking the means of safe milk production or storage. Households in the western world use it most often today for desserts and baking due to its unique flavour. It is also used as a substitute for pouring cream, as an accompaniment to desserts, or (undiluted) as a rich substitute for milk.

History

Condensed milk was introduced to the U.S. by Gail Borden which he made using a process under the patent issued on August 19, 1856. It became popular for those people who were remote from farm sources, since it was capable of long term storage.

The invention of evaporated milk followed three decades later when John B. Meyenberg emigrated to the U.S. from Switzerland where he had devised the process, but had no support to begin production. He obtained two U.S. patents for his process and sterilizing apparatus, issued on November 25, 1884. He formed the Helvetia Milk Condensing Company on February 14, 1885, with a number of farmers and businessmen of Highland, Illinois, as stockholders. By June 14, 1885, the first canned "Highland Evaporated Cream" was ready to be marketed. There were problems with the new product, with premature spoilage in early batches. Over the next few years, Louis Latzer and Dr. Werner Schmidt solved the problems which had been found to be caused by bacteria. With the marketing efforts of John Wilde, the company became successful as Pet, Inc., and is now part of The J.M. Smucker Co.

John P. Meyenberg, son of John B. Meyenberg, was the first American to evaporate goat's milk. He started the Meyenberg business in 1934, supplying goat milk products that are more digestible than cow's milk, and an alternative for people (like himself) who were allergic to cow's milk.

Definition

Evaporated milk is fresh, homogenized milk from which 60 percent of the water has been removed. After the water has been removed, the product is chilled, stabilized, packaged and sterilized. A slightly caramelized flavour results from the high heat process, and it is slightly darker in colour than fresh milk. The evaporation process also concentrates the nutrients and the food energy. Thus, for the same weight, undiluted evaporated milk contains more food energy than fresh milk.

In Malaysia

In Malaysia, due to price controls, evaporated (and condensed) milk contains palm oil. It is one of the ingredients to make *Teh Tarik* in Malaysia and Singapore.

Notable Producers

Evaporated milk is sold by several manufacturers:

- Carnation Evaporated Milk (the brand is now owned by Nestle and licensed to Smuckers in Canada)
- PET Evaporated Milk (now owned by Smuckers)
- Magnolia evaporated milk-(now produced by Eagle Family Foods)
- Nestlé evaporated milk
- F&N Evaporated Milk.

Scalded Milk

Scalded milk is milk that has been heated to 82°C/180°F. At this temperature, bacteria and enzymes in the milk are destroyed. Since most milk sold today is pasteurized, which accomplishes both of these goals, milk is typically scalded simply to increase its temperature. During scalding, a cooking utensil, a milk watcher, is used to prevent both boiling over and scorching(burning) of the milk.

Uses

- Scalded milk is called for in the original recipes for béchamel sauce, to prevent the sauce from thickening excessively. Since these early recipes predate pasteurization, this was a necessary step.
- Scalded milk is used in bread to make a more tender loaf.
- Scalded milk is used in yogurt to make the proteins unfold. The acid produced during the yogurt development causes less whey separation and a firmer yogurt.
- Café au lait, baked milk, and ryazhenka also use scalded milk.
- Scalded milk is used in many doughnut recipes.

Dried Milk or Milk Powder

Roller Dried Milk Powder

Dried milk (Milk Powder) products play a significant role in conserving the milk solids since their biological value can be retained for a long period of time under relatively simple storage conditions. The roller dried powder finds special application in the manufacture of infant foods, confectionery, ice cream, milk sweets etc.

Milk is concentrated by means of condensation to about 20-25 per cent of total solids in a vacuum pan or evaporator and fed to continuously revolving drum dricr, which are internally heated with steam. The film of the dried product is continuously scrapped off by a stationary knife or doctor blade, located opposite in the point of application of milk. The dried milk film thus obtained is ground, which is then sifted, packed and stored.

Method of Manufacture of Dried Milk

- The milk is separated (in case of skim milk powder) or clarified and standardized (in case of whole milk powder).
- The milk is fore-warmed to 95°C and re-condensed to 20% total solids. Homogenization of the condensed milk is done in case of manufacture of whole milk powder.
- The roller drums are observed for its correct alignment with the knife, cleanliness of the surface of the rollers, pressure build up in the drum and working condition of the powder conveyors and exhaust fan.
- The condensed milk in the feed line is taken in to the feed trough and the scraping blades are adjusted. The rate of milk flow is ensured to the determined level in the trough and the dried milk is scraped off from the surface of the drum uniformly.
- The semi dried and charred powders obtained initially have to be discarded. The conveyors are adjusted properly to carry the powder.
- When the milk flow is over, the conveyors are stopped and some hot water is run over the drums.
- The knife is taken off from the surface of the drum and supply of steam to the drum is cut.
- The rollers are stopped only when there is no steam pressure inside and the rollers are sufficiently cool.
- The vapour exhaust fan is stopped.

Spray Dried Milk Powder

Spray drying is the most important method of drying milk and milk products. By spraying in to a stream of hot air, the droplets formed present an extremely large amount of surface area and get dried immediately due to rapid evaporation of moisture. Milk is preheated and concentrated to 40-45% percent total solids. Hot air is filtered and directed in to the drying chamber. The concentrate is atomized to obtain small particles ranging from 10-100m in diameter. The air leaving the drying chamber enters a cyclone separator where the fine particles are collected. The dried products are cooled, sifted and packed in suitable packaging material.

Method of Manufacture of Spray Dried Milk Powder

- The milk is separated (in case of skim milk powder) or clarified and standardized (in case of whole milk powder).
- The milk is fore-warmed to 95°C and re-condensed to 40% total solids. Homogenization of the condensed milk is done in case of manufacture of whole milk powder.
- The spray drier is cleaned, pipelines are connected and general conditions of the valves and cyclone separators are checked.
- The steam is let in to the air heating coils. The exhaust fan is switched on, keeping the main door partially open.
- The blower is started and the main door is closed.
- The temperature of the hot air is adjusted to 180°C and this temperature is maintained for ten minutes to ensure the sterility of the unit.
- The atomizer is started and allowed to obtain the required speed.
- Initially, 20 litres of water is used for spraying through the feed pump.
- When the outlet air temperature reaches 100°C, the feed is changed to concentrate. The temperature of the inlet air entering the drying chamber usually ranges from 150-260°C and the outlet air exhausted from the drying chamber ranges from 100-105°C. the relative humidity of the drying chamber is quite low i.e. 3-4%.
- The *milk powder* is collected in the powder silo.
- When the flow of milk is over, the balance tank is flushed with about 20 litres of hot water (90°C) and the flow of milk to the dryer is reduced immediately.
- The atomizer and feed pump are stopped when the feed is empty.
- The steam supply to the air heating coils is stopped.
- The exhaust fan and air blower are stopped.
- The dried milk or milk powder is packed in required quantity in suitable containers.

10

Microbial Control by New Nonthermal Methods

Supermarket shelves today need to cater to the gourmet cook as well as the timeharried parent. Increasingly all types of consumers are demanding minimally-processed foods that are high in quality, nutritionally superior, and easy to prepare. Food processors have met this demand by developing refrigerated foods with extended shelf life. Ready-to-eat luncheon meats and complete heat-and-eat meals are some examples. By their very nature, however, these foods present challenges to ensure microbiological quality and safety. This Scientific Status Summary addresses the microbiological concerns associated with extended shelf life refrigerated foods and control measures for ensuring microbiological quality and safety.

Extended shelf life refrigerated foods are foods that have received minimal processing or precooking and have an enhanced but limited shelf life; refrigeration is a key preservation measure. These foods include conventional products, such as luncheon meats and cured meats, as well as a new generation of partially processed refrigerated foods (NFPA, 1988) such as meat, seafood, egg, and vegetable salads, fresh pasta and pasta sauces, other sauces, soups, entrees, complete meals, and uncured meat and poultry items. Sous-vide foods, cooked inside a hermetically sealed plastic package under vacuum, are also included in this definition.

If extended shelf life refrigerated foods are heat processed, the heat treatment is much less than that required for commercial sterility. Canned foods are, therefore, excluded from this food category. Sous-vide foods and others that receive a lower heat treatment than that used for canning and that require refrigeration are described by some authors as "refrigerated processed foods of extended durability" (Peck, 1997).

Microbiological Concerns

The chief microbiological concerns associated with these products centre around two types of microorganisms— psychrotrophic and mesophilic pathogens—that could grow during extended refrigerated storage or temperature abuse. Psychrotrophs are bacteria, yeasts, and molds that grow, although slowly, at refrigeration temperatures (below 7°C) but grow optimally at temperatures above refrigeration, e.g., 25–30°C. Their maximum growth temperatures are 30–35°C (Kraft, 1992; Olson and Nottingham, 1980). Mesophilic pathogens could survive under refrigeration and grow during any temperature abuse of the food. Mesophiles grow well between 20–45°C with optimum growth between 30–40°C (Jay, 1992). The potential for psychrotrophic spoilage microorganisms to grow during the extended refrigerated storage period and decrease organoleptic quality or spoil the food product is also a concern.

Pathogenic Microorganisms

Conventional wisdom of decades ago held that properly refrigerated foods would remain safe because it was thought that pathogenic bacteria could not grow at refrigeration temperatures.

Microbial growth was thought necessary to either produce a sufficient number of cells or enough toxin to cause foodborne illness. Since then, scientists have learned that several pathogens—such as Aeromonas hydrophila, nonproteolytic strains of Clostridium botulinum, Listeria spp., Yersinia enterocolitica, some strains of Bacillus cereus, enteropathogenic Escherichia coli, and Vibrio parahaemolyticus—can grow at refrigeration temperatures. Furthermore, scientists now know that some pathogens can cause illness when only a few cells are ingested. For example, as few as ten cells of the extremely virulent E. coli O157:H7 may cause hemorrhagic colitis (Buchanan and Doyle, 1997). Readers may refer to several texts for detailed information on the characteristics of pathogenic microorganisms and the foodborne illnesses they cause.

A. hydrophila is a facultative anaerobe that is generally considered a ubiquitous waterborne microorganism, occurring widely in fresh and brackish waters. Recent surveys (Pin et al., 1994; Saad et al., 1995; Schweizer et al., 1995) detected A. hydrophila in samples of raw milk, poultry, lamb, cheese, shellfish, pork, beef, watercress, lettuce, and escarole. Most cases of illness attributed to A. hydrophila have been sporadic, rather than associated with an outbreak (FDA, 1992).

C. botulinum is a ubiquitous anaerobic spore-forming bacterium whose spores are widely distributed in soil, freshwater and marine environments, raw agricultural products, and the intestinal tracts of fish and animals (Sugiyama, 1990). Four groups of C. botulinum (I-IV) and some strains of Clostridium baratii and Clostridium butyricum can produce botulinum neurotoxin (Hatheway, 1992). C. botulinum type I (proteolytic strains) and C. botulinum type II (nonproteolytic strains) are responsible for human foodborne botulism (Peck, 1997). The botulinum neurotoxins are differentiated as types A through G on the basis of serological reaction. The nonproteolytic strains—type E and some type B and F—do not produce overt signs of food spoilage during growth and toxin production.

Some nonproteolytic strains of C. botulinum are a concern with extended shelf life refrigerated foods because with sufficient time they may be able to grow and produce neurotoxin at temperatures as low as 3.3°C. Proteolytic strains, which are mesophilic, may be able to grow and produce toxin in foods if temperature abuse occurs. Most outbreaks of botulism in the United States have been caused by home-processed vegetables, fish, or meat products (ICMSF, 1996). The incidence of botulism from consumption of refrigerated foods is exceedingly low. However, the few outbreaks that have occurred and research challenge studies illustrate the potential C. botulinum hazards associated with extended shelf life refrigerated foods (Conner et al., 1989).

L. monocytogenes, a facultative anaerobe, is ubiquitous in the environment. L. monocytogenes has been isolated from soil, silage, food processing environments, and healthy humans and animals (ICMSF, 1996). A variety of foods, such as refrigerated ready-to-eat meat sandwiches and meat salads have been recalled from the marketplace because of contamination with L. monocytogenes (Ryser and Marth, 1991). Individuals with compromised immune systems, e.g., newborns, the elderly, and people suffering from the acquired immunodeficiency syndrome, are most susceptible to listeriosis. Outbreaks of listeriosis in North America have been associated with coleslaw, soft Mexican-style cheese, and milk (McLauchlin, 1996).

Y. enterocolitica is a facultative anaerobe whose main reservoir of bioserotypes pathogenic to humans is believed to be the pig (ICMSF, 1996). Symptoms of yersiniosis, the disease caused by Y. enterocolitica, may include fever, diarrhea, headache, vomiting, and severe abdominal pain similar to that associated with appendicitis. Y. enterocolitica has been isolated from a variety of animals, foods (lamb, pork, oysters, shrimp, and crabs), and water; however, isolates are often avirulent. Outbreaks of yersiniosis, which are relatively uncommon in the United States, have been caused by contaminated chocolate milk, recontaminated pasteurized milk, bean sprouts, tofu, and chitterlings (raw pork intestine).

B. cereus, an aerobic spore-former including psychrotrophic and mesophilic strains, is widely distributed in nature and in foods. B. cereus is commonly found in soil, milk, cereals, starches, herbs, spices, and other dried food products and on the surfaces of meats and poultry. B. cereus can produce two toxins that cause two distinct types of illness—a diarrheal illness and an emetic illness characterized by nausea and vomiting. Every well-documented report of B. cereus intoxication has described time and temperature abuse that enabled initially relatively low (innocuous) levels of B. cereus in foods to increase greatly. In most incidents, the food vehicle was a cereal or cereal- or spice-containing product (ICMSF, 1996).

A few serotypes and strains of E. coli, a facultative anaerobe that is part of the normal microflora of the intestinal tract of humans and most warm-blooded animals, can cause illness. Although they are not considered true psychrotrophs, some of these strains can grow at 6.9°C and below. Pathogenic E. coli are categorized into six groups—enteropathogenic, enteroinvasive, enterotoxigenic, enterohemorrhagic (EHEC), enteroaggregative, and diffusely adherent.

Foods involved in outbreaks caused by pathogenic E. coli include meat, poultry, fish, vegetables, apple cider, raw milk, Brie and Camembert cheese, water, and radish and alfalfa sprouts. Some strains of E. coli, including some EHEC strains, are acid tolerant, a complex phenomenon that is growth phase dependent and inducible; acid tolerance may persist for extended periods at refrigeration.

V. parahaemolyticus is a facultatively anaerobic halophile (requiring sodium chloride for growth) occurring worldwide in inshore marine waters and frequently associated with molluscs, crustaceans, and fish (ICMSF, 1996). Although the microorganism is considered mesophilic, growth has been demonstrated at temperatures as low as 5°C (Twedt, 1989).

The microorganism is the most common cause of foodborne illness in Japan because it frequently contaminates seafood, often eaten raw in that country. Contaminated raw, improperly cooked, and cooked recontaminated fish and shellfish have been implicated in cases of gastroenteritis.

Spoilage Microorganisms

With sufficient time at refrigeration temperatures, several types of psychrotrophic bacteria, yeasts, and molds may grow to levels sufficient to cause food spoilage. These microorganisms include the Acinetobacter- Moraxella group, Alcaligenes species, Flavobacterium spp., Microbacterium spp., Xanthomonas spp., and the microorganisms of primary concern in extended shelf life refrigerated foods: Brochothrix thermosphacta, lactic acid bacteria (LAB), and Pseudomonas spp.

B. thermosphacta, which is aerobic (requiring free oxygen) to facultatively anaerobic (growing well either aerobically or anaerobically), has been recovered from vacuum-packaged beef, pork, lamb, and heat-processed cured meats such as sliced cooked ham, corned beef, and others (Kraft, 1992). The extent of spoilage of vacuum-packed meat by B. thermosphacta varies with product pH and with oxygen permeability of the packaging material (Egan and Grau, 1981). Spoilage may involve development of sliminess and production of off-odors and off-flavours conferred by short chain fatty acids (Jay, 1992).

The LAB, such as Lactobacillus spp., Leuconostoc, and Pediococcus, are facultatively anaerobic. The type of spoilage produced by the LAB is determined by the nature of the bacteria. Homofermentative LAB produce primarily lactic acid during sugar fermentation. Heterofermentative LAB produce acetic and formic acid, ethanol, and carbon dioxide, in addition to lactic acid. LAB can spoil a variety of foods, including milk and milk products, meats, vegetables, fruit juices, sugary products, alcoholic beverages, and products preserved with vinegar.

Pseudomonas spp., which are aerobic, are among the most common spoilage agents of refrigerated foods. Growth of Pseudomonas spp., like that of other Gram-negative psychrotrophs, is affected by oxygen tension, salt and other food additives, water activity (aw), pH, and other factors. During growth, pseudomonads produce proteases and lipases that can catalyze reactions causing degradation of protein and fat. The consequence of these reactions is formation of peptides and fatty acids of undesirable flavour (e.g., bitterness, rancidity) and odour. Sometimes these bacteria also produce unsightly green pigments.

Many yeasts and molds given sufficient time under refrigeration temperatures can spoil fruit juices, meat products, vegetables, dairy products, and possibly other foods. Some yeasts in the genera Candida, Hanseniaspora, and Saccharomyces can grow in fruit juices at -5.5°C to -2.2°C, just above the freezing temperature for these foods (Pederson et al., 1959). Several genera of yeasts are found on fish and meat products. These include Candida, Cryptococcus, Debaromyces, Hansenula, Pichia, Rhodotorula, Saccharomyces, Sporobolomyces, Torula, Torulopsis, and Trichospora (Jay, 1987). Growth of yeasts on foods is commonly accompanied by production of carbon dioxide and yeasty, fruity, or alcoholic off-flavours and odors.

Psychrotrophic molds include Botrytis cinerea, Geotrichium candidum, Pullaria pullulans (Aureobasidium pullulans), and some species in the genera Alternaria, Monilia, Mucor, Penicillium, Sporotrichium, and Rhizopus. Not only does the visible presence of mold indicate spoilage, molds also commonly produce enzymes that degrade carbohydrates, fats, and proteins, causing softening of foods and flavour and aroma deterioration. Some species of molds, especially those in the genera Aspergillus, Fusarium, and Penicillium, can produce mycotoxins. Aspergillus spp. cannot produce mycotoxins at refrigeration temperatures, whereas certain species of Fusarium and Penicillium can (Frisvad and Samson, 1991).

Control Measures

Several types of control methods are effective in preventing or minimizing microbial contamination of product and inhibiting the growth of or destroying microbial contaminants.

Good Manufacturing Practices

(GMPs), Sanitation, and Hygiene. Processors need to select high-quality raw materials with low levels of microorganisms, especially psychrotrophs. They need to determine potential microbiological hazards of ingredients, possibly using microbiological specifications for ingredients to minimize risk (Moberg, 1989).

Fabrication of raw materials into finished products under hygienic conditions is also important. Food processing equipment must be designed and constructed so that it: (1) is inert to the product, (2) has smooth and nonporous product-contact surfaces, (3) is readily accessible for cleaning and inspection, (4) is self-emptying or self-draining, (5) has covers to prevent external contamination, and (6) has readily cleanable surfaces that do not contact the product and do not harbour contaminants (Cliver and Marth, 1990). The equipment should be cleaned and sanitized as often as is necessary during a day's operation to prevent development of a biofilm that can contaminate subsequent lots of product. Cleaning and sanitizing adequacy can be determined using the more traditional swab procedures, the RODAC (agar contact) method, or the newer rapid ATP (adenosine triphosphate) bioluminescence assays.

Filtration of air entering the food processing area reduces the number of airborne contaminants. If processed foods will not receive a heat treatment or will have few barriers to microbial growth, use of "absolute" (high efficiency) air filters can virtually eliminate microbial contamination. If an air conditioning system is present, the system must be maintained properly so that condensate drains freely and does not contaminate the product. Food processing personnel must use hygienic practices and must be barred from moving from areas containing raw materials to areas containing finished products.

GMPs, sanitation, and hygiene are necessary prerequisites for implementing an effective Hazard Analysis Critical Control Point (HACCP) system, which enables the highest level of food safety assurance possible. HACCP is a systematic approach to the identification, evaluation, and control of food safety hazards, from raw material production and procurement to distribution and consumption of the finished product (NACMCF, 1997). HACCP is based on seven principles: (1) conduct a hazard analysis, (2) determine the critical control points, (3) establish critical limits, (4) establish monitoring procedures, (5) establish corrective actions, (6) establish verification procedures, and (7) establish record keeping and documentation procedures.

Multiple Barriers/Hurdles

Referred to as the hurdle concept or hurdle technology, this approach combines several factors at subinhibitory concentrations that can effectively control microorganisms in refrigerated foods. Common hurdles include physical elements—such as refrigeration, modified atmosphere packaging (MAP), and heat treatment— and physicochemical factors—such as aw, pH, and preservatives. When used together, hurdles interact, sometimes synergistically, enabling use of lower intensities of each factor than would be necessary if each were used alone.

Scott (1989) recommended that challenge studies be conducted to verify the effectiveness of the combination of hurdles. For example, Simpson et al. (1995) demonstrated the antibotulinal effects of salt and pH in minimally processed sous vide spaghetti and meat sauce on proteolytic C. botulinum types A and B spores and toxin production. The challenge studies indicated that the probability of toxigenesis increased with storage time, but that it decreased as either the aw or pH was decreased. Growth of the spores and toxin production were prevented in the product that was processed at 75°C for 36 min and held at 15°C (simulating mild temperature abuse) for 42 days by >1.5% salt (aw 0.983) and pH 5.5.

Ingredients

Some products can be formulated with ingredients that are barriers to microbial growth. For example, organic acids, particularly acetic but also lactic and citric, can inhibit bacterial growth. Sorbate, propionate, and benzoate have both antibacterial and antifungal properties. Although salt has antimicrobial properties it is not commonly used in high enough concentrations to be effective; this is particularly true of "low-sodium" foods. When appropriate, use of salt and other ingredients to reduce the aw to 0.980 or below will lengthen the lag phase of most bacteria and will further reduce the rate of any subsequent growth.

Heat Treatment

Heating foods will reduce the microbial population; the degree of reduction depends on the magnitude of the heat treatment, i.e., time and temperature. The magnitude of heat treatments commonly used is pasteurizing (destructive to vegetative pathogens) rather than sterilizing. Heat treatments lower than those producing commercial sterility are likely to inactivate vegetative cells, but not bacterial spores. Once the heating is completed, stringent hygienic measures must be implemented to prevent recontamination of the food with psychrotrophic spoilage or pathogenic microbes.

Modified Atmosphere Packaging

MAP extends product shelf life by reducing oxygen and/or increasing gases, such as carbon dioxide, in the food product environment. MAP inhibits the growth of aerobic spoilage microorganisms, such as Pseudomonas species, but allows facultative anaerobes such as LAB to grow. Integrated with aseptic packaging, the technology has experienced rapid growth in the minimally processed refrigerated foods sector (Brody, 1996).

Use of MAP is not without some risk, however. Any facultatively anaerobic or anaerobic psychrotrophic pathogens, such as nonproteolytic C. botulinum type E or Y. enterocolitica, may be able to grow until the LAB have reduced the pH of the product to inhibitory levels. Further, unlike aerobic spoilage microorganisms, growth of LAB may not be accompanied by overt evidence of spoilage. Moreover, if the MAP product received a nonsterilizing heat treatment, any surviving C. botulinum spores may, upon temperature abuse of the product, germinate, grow, and produce toxin without organoleptic signs of spoilage.

Storage Temperature and Shelf Life

Microbial lag phases (during which there is no growth or a decline in microbial numbers) and generation times (duration between formation of a daughter cell and its division into two new cells) increase as refrigeration temperature decreases. Thus, product temperature should be maintained just slightly above freezing. Acceptable product shelf life (e.g., days or weeks) at specified temperature limits should be established and monitored to help manage food quality and safety. Because the potential exists for temperature abuse at some point during handling or for storage past the intended shelf life, timetemperature indicators or integrators can be useful in determining when refrigeration temperatures or intended storage times have been exceeded.

Other Control Measures

Except for the limited use of food irradiation, the bacteriocin nisin, and high hydrostatic pressure (HHP), these methods are not yet fully developed nor commercially applied.

Ionizing radiation, from gamma rays (produced by the radioisotopes cobalt-60 or cesium-137), machine generated x-rays (with a minimum energy of 5 million electron volts, MeV), and electrons (with a maximum energy of 10 MeV) has been studied extensively.

The United States has accepted this nonthermal processing technology for insect disinfestation of wheat, wheat flour, and fresh fruits and vegetables, inhibition of maturation of fresh fruits and vegetables, sprout inhibition of potatoes, inactivation of Trichinella spiralis in pork, and microbial decontamination of spices, herbs, vegetable seasonings, poultry, and red meats. Commercial application of ionizing radiation to foods in the United States, however, has grown slowly (Olson, 1998). Widespread application to refrigerated foods requires consumer acceptance.

Pulsed electric fields technology (PEF), also nonthermal, uses very short pulses of high intensity electric fields to inactive microorganisms. It has been largely limited to liquids such as juices, milk, and liquid egg. Applied to foods, PEF has the potential to equal conventional pasteurization (Yousef, 1996).

Pulsed high-intensity light is a nonthermal technology that uses a xenon flashlamp to generate extremely brief (< 2 msec) flashes of intense broad-spectrum (200 to 1,100 nanometer wavelength) light to inactivate microorganisms (yousef, 1996). accepted by the food and drug administration (fda, 1996) for controlling microorganisms on the surface of food, the technology is also useful for treating surfaces of equipment and packaging materials. successful commercial application to food requires further development because the method as currently understood suffers from limited penetration into food and may cause lipid oxidation (yousef, 1996).>

HHP is effective in controlling microorganisms. Raffalli et al. (1994) demonstrated that L. innocua added to 35% fat cream at 107 cells per milliliter was reduced 98.7–99.99% after treatment at 25–26°C for 10–30 min. The D-value for L. innocua was 7.4 minutes. After treatment for 20 and 30 min, all surviving microorganisms were injured; a resuscitation step was needed before they were able to grow on a selective medium. HHP is applied commercially to refrigerated avocado products.

Bacteriocins are antimicrobial proteins produced by certain bacteria. The best known bacteriocin is nisin (also designated as a lantibiotic), produced by certain strains of Lactococcus lactis subsp. lactis (formerly Streptococcus lactis). In the United States, nisin is generally recognized as safe (GRAS) for limited use in pasteurized processed cheese to control growth of and toxin production by C. botulinum (FDA, 1988). GRAS petitions have been filed for use of nisin in reduced cholesterol liquid whole eggs, sauces, and nonstandardized salad dressings (FDA, 1994, 1995). In recent years, an array of other bacteriocins, many of which are inhibitory to foodborne pathogens, has been discovered. These include unnamed bacteriocins produced by enterococci, pediocin produced by Pedicoccus acidilactici (Huang et al., 1994), bavaricin produced by Lactobacillus bavaricus, mesenterocin produced by Leuconostoc mesenteroides, carnocin produced by Carnobacterium piscicola (Bagi and Buchanan, 1994; Mathieu et al., 1994), sakacin produced by Lactobacillus sake (Holck et al., 1994), and curvaticin produced by Lactobacillus curvatus (Garver and Muriana, 1994).

Labeling

Manufacturers recognize the potential for temperature abuse during distribution or storage of foods requiring refrigeration. Hence, they voluntarily use label statements, such as "keep refrigerated" or "refrigerate after opening," to inform consumers of the need to maintain product at refrigeration temperatures (FDA, 1997).

The FDA determined in 1997 that the labeling of potentially hazardous foods that need refrigeration by consumers should be more specific about the types of hazards present and the necessary storage conditions after the food is opened and issued labeling guidance (FDA, 1997) to food manufacturers. The agency delineated foods that need refrigeration into three groups and developed model label statements and guidance on label placement and prominence. The model label statements refer to the importance of refrigeration for foods in two of the groups to maintain safety and the use of refrigeration for foods in the third group to maintain quality.

Forgive the pun, but there are some killer microbial intervention technologies on the market that are of keen interest to the food industry. The number of commercially available thermal, non-thermal, antimicrobial chemical and other novel treatments applied to foods and/or food-contact surfaces designed to inhibit, reduce or inactivate microorganisms has never been greater—or more in demand by food manufacturing operations.

What has driven a surge in the development and availability of new and improved interventions? Certainly, part of the answer is that regulatory, industry and research institutions have formed a strong partnership supporting the use of quantitative microbial risk assessments to identify critical points within the food chain at which effective interventions will have the greatest impact on reducing foodborne illness. Several of the U.S. Food and Drug Administration's (FDA) and the U.S. Department of Agriculture's (USDA) more recent industry guidelines, directives and proposed rules reflect this unified push to apply intervention technologies to improve food safety. FDA's 2004 Produce Safety Action Plan, for example, identifies as part of the agency's objectives the promotion of risk assessments to identify with industry the microbial interventions that will be most effective in produce operations.

Similarly, USDA's Food Safety and Inspection Service (FSIS) interim final rule on the control of *Listeria monocytogenes* in ready-to-eat (RTE) meat and poultry products (Directive 10,240.4) provides incentives for RTE producers to use post-lethality treatments, antimicrobials ingredients at formulation and other intervention technologies to significantly reduce the risk of the presence or growth of *Listeria* on these products. And, industry has been vocal in expressing its desire that new intervention strategies receive expedited regulatory agency review, as companies search for viable options that will improve their ability to prevent, control and eliminate undesirable microorganisms through HACCP, GMP and other food safety programs.

As *Food Safety Magazine's* editors have traveled the country this year, we've been introduced to several innovative, new or improved technologies and systems designed for use as part of the multi-hurdle intervention approachto food safety. Here, we'll highlight some of these in-plant thermal, non-thermal, chemical and novel microbial intervention systems as a quick guide to the latest advances in this important area of food safety.

Hot Thermal Interventions

Applying heat to foods and food-contact surfaces, either as a validated kill step or as a pathogen reduction method, is the classic means by which to achieve processing, packaging and post-packaging intervention. From steam treatments to dry heat to cook steps, the higher the temperatures applied, the higher the confidence that bacteria have been destroyed. For some foods, however, heat treatments can have an adverse affect on the quality attributes of finished product, such as minimally processed fruits and vegetables. Some of the newer microbial intervention technologies are designed with this issue in mind and thus are capturing the attention of food manufacturers nationwide.

One such thermal technology, Biosteam Technologies, Inc.'s trademarked ThermoSafe system, is a pasteurization method that has proven effective in the area of surface decontamination of fruits and vegetables destined for further processing, including fresh juice, cut fruit and melons, fruit concentrates and smoothies. The company reports that research conducted by Food Safety Net Services, Ltd., an ISO/IEC 17025 accredited network of laboratories, shows that the ThermoSafe process can be effective in reducing the microbial load of *Listeria monocytogenes, Salmonella* spp., *E. coli* O157:H7, and *Lactobacillus* spp. on the surface of fruits and vegetables by 5 logs, and in some cases, up to 9 logs. In addition to the safety factor of the process, studies indicate taste panels were able to determine a sweeter flavour from the steamed fruit over the control or unsteamed product.

In describing the process, Biosteam notes that the Thermo Safe process uses time and temperature settings, depending on the commodity, to facilitate the desired microbial intervention step. The system can serve to clean and sanitize both products and equipment in minutes, automatically ending the sanitizing cycle at a predetermined temperature. This technology is designed to provide the user with a constant supply of steam, whether in the form of a portable unit or a large scale in-line applications system.

Similarly, ALKAR-RapidPak's Surface Pasteurization Technology (RP-SP) is designed to produce safer food products without loss of desirable product sensory characteristics. This flash pasteurization system promises a 3-log reduction for Listeria monocytogenes, for example, with no formula change required and no effect on packaging films. While a conventional pasteurization method essentially recooks product, taking between four to 20 minutes of application time that must be followed by re-chilling, the RP-SP is designed to achieve surface pasteurization in 1.5 seconds. The system module uses bursts of steam and a jostling mechanism to pasteurize all surfaces of food products, followed by the injection of rapid bursts of high-pressure steam.

"Cool" Interventions Emerge

Once considered "novel" technologies relegated to pilot plant research, non-thermal intervention systems are beginning to take their place in the real world of the food plant. A few of these technologies have been available to industry for awhile, but have recently experienced a resurgence of interest as application ranges have broadened and procedures for use are better understood by end-users. For example, since obtaining FDA GRAS status for use as a direct food contact antimicrobial agent in 2001, ozone is gaining many industry adherents as an additional intervention to strengthen food safety programs, such as HACCP and GMPs, in the processing plant and beyond. Ozone, which is an effective biocide against biofilms, bacteria, viruses, fungi and protozoa, has long been successfully used for no-rinse sanitation of food-contact surfaces and in clean-in-place (CIP) systems in food plants. Today, ozone also can be dissolved into water and used for direct contact on fruits and vegetables, raw and ready-to-eat meat and poultry, fish and commercial eggs, making this an intervention that can be used along the entire food production chain.

According to DEL Ozone, which manufactures the AGW Series of ozone-enriched water systems for sanitation and direct food contact applications, the primary benefit of ozone-enriched water is that it kills microbes as effectively as chlorine, sans residual chemicals. For example, DEL's ozone-enriched water efficacy testing, performed by the National Sanitation Foundation (NSF) under AOAC International methodology, shows that ozone dissolved in water can provide a 6-log reduction of *Staphyloccoccus aureus, Salmonella choleraesuis* and *Pseudomonas aeruginosa* in as little as 30 seconds (depending on the microorganism). The testing also shows that a 5-log reduction of *E. coli*, and a 4-log reduction of both *Listeria monocytogenes* and Campylobacter jejuni can be obtained.

DEL Ozone systems enable operators to apply ozone in both aqueous and gaseous forms to fruits, vegetables, meat, poultry and fish products, and can be easily integrated into flume, cascade, or drench type operations, as well as into both ozone prewash and ozone wash sections.

Ozone's effectiveness as an intervention goes beyond pathogens. In studies conducted by Silliker, Inc., Novazone patented ozone-based purification technologies have been shown to be effective in disinfecting, inhibiting or removing mold growth, reducing contamination and decay, and improving shelf life. In one study, Novazone applied gaseous ozone for one night in a 175,000-cubic-ft. cold storage room containing contaminated lemons with a measurable aerobic plate count of more than 3 million colony forming units (CFUs) of microorganisms per gram. The independent testing showed that the ozone treatment not only reduced contaminants by a 2-3 log count, but in the bin dump room, also reduced the yeast count from 900,000 CFUs (untreated) to 32,000 CFUs (treated) and the mold count from 400,000 CFUs (untreated) to 70,000 CFUs (treated).

Again, non-thermal intervention technologies promise the food industry effective pathogen reduction as well as enhanced product quality and shelf life. Ozone technology is no exception, according to The BOC Group, whose ozone-based pathogen control technology from its food safety technology and consulting division, Intervent, is receiving good response from clients such as the Norway-based Stolt Sea Farm, a $1 billion global aquaculture company that has installed the technology in its Los Angeles, CA and Stratford, CT plants. The plants, which process sturgeon and salmon, have made the Intervent system a core component of their pathogen reduction strategies, and report that validation tests have confirmed that, in addition to reducing pathogens, ozone had a significant, positive impact on product quality and shelf life. As a result, Stolt plans to eventually introduce the technology in all of its processing plants.

Continued research also has informed system design improvements to ensure the delivery of the correct amount of ozone reliably and safely to food products and food-contact surfaces to increase the intervention benefit. Many ozone technology developers, for example, now have in-depth material compatibility sheets to increase the ensure the good condition of equipment and components after treatment. While aqueous ozone is compatible with many materials, including ceramic, glass, silicone and Teflon seals, and stainless steel piping, in the recent past, some end-users were unaware that aqueous ozone should not be applied on surfaces made with natural rubbers, polyurethane or resin-based plastics.

In an effort to provide informative materials on the technology's proper application, a new player in the market, Ozone Safe Food/eFoodSafety.com, has launched an in-depth website to help industry enhance its understanding of this intervention. The company offers several equipment models that feature its patented Vortex Mixing System, which compresses the ozone into microscopic bubbles in the water solution. The result, says the company, is higher ozone residuals with less off-gas to achieve a faster, more effective application.

High pressure processing (HPP), an automated system that kills microorganisms by interrupting their cellular function without the use of heat, is another intervention that is becoming more widely available—and affordable—for food processors. HPP emerged in the 1990s as a food pasteurization alternative to heat or irradiation processing but initial costs of the systems have prevented their widespread use. Studies show that the process also extends product shelf life by inactivating spoilage organisms, and when appropriately used, does not alter the texture, appearance or flavour of foods. HPP is effective on food products such as RTE meat, cheese, guacamole, fruit juice, onions, salsa, oysters and other seafood.

In addition, in preliminary studies reported in March 2005, FDA/USDA research collaborators found that one-minute HPP treatments of oysters significantly reduced hepatitis A virus from study samples. Researchers also indicated that the technology might be successfully used as an intervention against noroviruses in the future.

As the first commercial tolling processor, American Pasteurization Co. (APC) has introduced what it describes as access to food processors with volumes too low to justify the investment in HPP equipment. The company works with processors to incorporate HPP into their current distribution chain, so that food is not only treated at APC, but can be date-stamped, labelled, packaged and boxed on its way to distributors and retailers. APC aims to make HPP easier and more convenient by eliminating extra shipping steps for small- to mid-sized processing operations that are interested in supplementing their existing manufacturing practices and processes with an additional food safety intervention step.

Avure Technologies Inc. recently introduced its next-generation 215L ULTRA HPP system for higher-volume plant operations, particularly for sliced RTE meat processors looking for a post-packaging lethality step that inactivates Listeria monocytogenes in accordance with the recent USDA FSIS directive. This latest design upgrades Avure's 215-liter pressure vessel with a dual-pump pressurization system that uses a new generation of pressure intensifiers to accelerate process steps. Along with an advanced PC-based control system that allows multiple units to be monitored and controlled from a central location, Avure also has shortened the system's cycle times to improve throughput.

Beginning in 1990, FDA approved irradiation (ionizing radiation) treatments as a safe and effective microbial reduction method for categories including spices, poultry and eggs, red meats, ground beef, seafood, sprouts, and fruits and vegetables. Currently, FDA is considering adding RTE meats, including hot dogs, deli ham and turkey and bologna, to the list of foods that can be treated with ionizing radiation. Earlier this year, USDA researchers with the Agricultural Research Service (ARS) Food Safety Intervention Technologies Research Unit reported that there may be other benefits to using this non-thermal method, which inactivates pathogens, parasites and spoilage organisms by destroying or damaging their cell membranes.

In studies on the use of irradiation to eliminate *Listeria monocytogenes* from RTE meats, the USDA researchers found that certain common food additives, such as a combination of acetic acid salts and lactic acid in bologna formulations, make *Listeria* more sensitive to radiation, thus allowing a significant reduction the dose of radiation needed to inactivate the pathogen to maintain product quality attributes.

The researchers noted that the combined treatments also prevented growth of spoilage microorganisms for two months in some RTE meats. "Ionizing radiation, when combined with common food additives, has the potential to significantly reduce the incidence of listeriosis associated with consumption of RTE meats in the U.S.," concluded the scientists.

Commercially available irradiation technologies, such as gamma processing technology pioneer MDS Nordion's Centurion and Sterigenics' electron beam, gamma and X-ray processing technologies are well established as effective microbial interventions, from receiving through processing to post-packaging applications. Given the findings of this new research, the food industry will likely see some enhancements of irradiation technologies and systems to incorporate additional things like food additives.

Chemical Interventions on the Rise

One of the most active areas of new technology development is chemical intervention systems, in the form of both direct food additives and secondary additives, or food contact-surface treatments.

Mionix's Safe2O brand line of products makes the list, with both USDA approved, multi-platform pasteurization technologies and proprietary antimicrobial agent formulations. The Safe2O RTE 03 chemical intervention, for example, involves a unique blend of acidic calcium sulfate, lactic acid and propionic acid that kills bacteria on contact and prevents further growth of *Listeria monocytogenes* on various ham products and other RTE food products. The product is a validated one-step solution to achieve Alternative 1 in FSIS Directive 10,240.4.

Surface-friendly chemical solutions, such as peracetic or peroxyacetic acids, also are gaining popularity. These microbial interventions are compatible with a range of surface materials, providing antimicrobial efficacy as well as protection from corrosion or other chemical-caused damage to equipment. For example, AFCO's PeraSafe, composed of 15% peracetic acid and 10% H2O2, combines the proven efficacy of an oxidizer with an organic acid, reducing corrosion rates on machinery. PeraSafe is used as a direct microbial intervention on meat and poultry carcasses and in fabrication applications, including ground beef, cuts and trims.

Ecolab's EcoShield, its integrated interventions systems program that offers the food industry a holistic, multiple-intervention approach, combines antimicrobial chemical treatments, irradiation technology and sanitation control systems to reduce or eliminate pathogens. Among its chemical antimicrobials, Ecolab's Inspexx 100 and 200 peroxyacetic acid-based food surface treatments are recommended for use in poultry spray rinsing and chill water to reduce microbial contamination on processed poultry surfaces, and for use in carcass sprays to control microbes on red meat surfaces, respectively. When used with certain dispensing equipment, notes the company, system performance is optimized and greater antimicrobial efficacy against Salmonella typhimurium, Campylobacter jejuni, Listeria monocytogenes and *E. coli* O157:H7 is achieved.

Ecolab also offers Tsunami, a peroxyacetic acid-based additive used for microbial control in fresh-cut, processed and post-harvest fruit and vegetable flume and wash water systems. PURAC America's Purasal Opti.Form, a blend of sodium or potassium lactate and sodium diacetate, also is well-suited for the multi-hurdle intervention approach. Research by Oscar Mayer's R&D team has shown that the antimicrobial can effectively suppress the growth of Listeria monocytogenes in RTE meats.

More Interventions of Interest

Other novel or improved technologies and applications of interest to food processors provide the opportunity for intervention and reduction of overall bioburden as a way to further enhance food safety. For example, e-Merge Interactive's fluorescence-based real-time imaging technology, VerifEYE, which is a used by meat processors in the U.S. to instantly detect and verify trace amounts of organic contamination on meat surfaces that can harbour pathogens such as *E. coli* O157:H7 and Salmonella, provides a detailed display of surface contamination of incoming meat products, which can then be accurately and rapidly removed—before carcasses travel further into the processing plant. The technology, which is available in a portable handheld version called the Solo, designed for spot-checking surfaces, as well as in an automated, in-line Carcass Inspection System (CIS), for automated inspection of entire carcasses, has been recognized as an effective tool for reducing contamination prior to the application of other microbial interventions during further processing.

Engelhard Corp. is building the antimicrobial agent directly into packaging film in an innovative intervention application. The company's Aseptrol antimicrobial, a chlorine dioxide formulation that rapidly kills bacteria, viruses and cysts, is being incorporated into antimicrobial films that will enhance long-term or residual control of microbial contamination, as well as potentially extend shelf life. According to Engelhard, applications for these antimicrobial films range from packaging meat, poultry and seafood, to fresh fruits, vegetables and foods in transit or storage from producers to processors or institutional users.

Similarly, Next Conveyor (formerly Materials Systems Engineering) has built a preventive barrier to bacterial bioburden directly into its NEXGEN conveyors. The company incorporates antimicrobial belting and bacteria-killing components as part of its sanitary design of NEXGEN conveyors. By using Westlake Plastics Technologies' Silveron product, the patented silver-based antimicrobial technology belonging to Milliken Chemical, the next-generation conveyors are kept cleaner by preventing or reducing the formation of biofilm on the equipment surface. The silver ions provide the product with its broad antimicrobial characteristics and are contained in tiny ceramic crystals that release the silver as needed. The controlled release and dispersion of the additive throughout the components ensures long-term antimicrobial intervention.

AgION Technologies' AgION antimicrobial coating technology another silver ion delivery system being used in the manufacture of building and equipment components to inhibit microbial growth. Interesting applications include the use of AgION in hardware and plumbing used in the foodservice industry, and in commercial HVAC air handling units.

Julie Larson Bricher is Editorial Director of Food Safety Magazine.

KSU Process Protects Ready-to-Eat Beef from Pathogens

The good news for consumers is that there are more ready-to-eat (RTE) meat products on the market and that vacuum packaging has made it possible to keep them fresh in appearance and taste. The problem is that pathogenic bacteria can grow on the meat in this packaging at both room and refrigeration temperatures.

The better news is that Food Safety Consortium researchers at Kansas State University have found that a solution of sodium citrate can inhibit the growth of the bacterium *Clostridium perfringens* on restructured roast beef.

RTE meats go through a mild heat treatment, but the treatment stimulates rather than reduces the growth of the bacterium in vacuum packaging. One way to prevent the problem would be to follow federal guidelines to sharply cool down the meat within five hours. But not all of the current refrigeration technology makes that possible.

So, according to KSU food science professor Dr. Daniel Y.C. Fung, "There is a need for additional secondary safety barriers in vacuum-packaged meat products that will help prevent the growth of anaerobic bacteria such as C. perfringens during cooling procedures."

C. perfringens is a common foodborne bacterium that the federal government reported was responsible for more than 6% of bacterial foodborne disease outbreaks in 2000. Its significance lies also in its designation by the Centers for Disease Control and Prevention as an organism that can produce toxin that could potentially be used in a bioterrorist attack.

Microbial Genomics: A New Tool to Increase Food Quality and Safety

There's a new discipline in town: genomics, a new field of science that analyzes and compares the complete genome (genetic material of an organism) of organisms or a large number of genes in a simultaneous fashion. When the media commented on the success of the human genome project, everybody expected genomics to greatly improve medicine. Now genomics is also entering food production and processing. Microorganisms play important roles in our foods. Microbial Genomics can help us understand what microorganisms do and how they do it, in ways that were not previously possible, helping us to better understand how they can be manipulated for our benefit. Future foods will benefit from more efficient, less costly processing methods, better quality, enhanced freshness and longer shelf lives.

Consumer demand for fresh tasting products and convenience foods - such as ready to eat chilled multi-component products - is increasing rapidly. Therefore, processing methods for mild preservation and novel, non-thermal preservation technologies are increasingly needed.

However, a major limitation of classic food microbiology research is that the effect of preservation strategies on spoilage microorganisms can only be determined after a number of days and with separate determinations for each organism. This classic approach is very time-consuming since it consists in tasks such as colony counting of surviving microorganisms or toxin production on plates.

Hence, it is important to develop better knowledge and methods, enabling a quick and reliable prediction of the best processing conditions and a direct insight in the effectiveness of the conditions being applied.

Microbial Genomics in Food Production and Processing

Genomic technologies offer a new alternative to the classic approach. They allow quick identification of microorganisms present in the (raw) product and they could help to directly measure the total response of the target spoilage microorganisms to the applied preservation methods. By making use of this new approach, known as applied microbial genomics, it might be possible one day to reduce the number of experiments needed to measure all relevant responses.

The tools used for that are small chips containing the information of thousands of genes of food spoilage micro-organisms, which are attached to a solid surface (like a glass slide) in a grid-like array. These so called "microarrays" could enable in the long run the outcome prediction of a preservation treatment and the definition of additional preservation steps if necessary. As a result, process control could be improved significantly and the energy input into preservation processes could be reduced. This is hoped to result in improved sensory properties and significant energy savings (due to tailor-made process conditions) and decreased product losses.

An Example: Campylobacter Jejuni

A number of Campylobacter species, grouped under the name of thermophilic campylobacters, are "formidable" pathogens. Among them, Campylobacter jejuni is one of the world's most "successful" food poisoning bacteria and is probably responsible for more than twice as many cases of poisoning as Salmonella. Until recently, however, there has been relatively little research into why the bacterium is so virulent, but work in the UK is beginning to shed some light on the problem.

C. jejuni is thought to have 1,700 genes. Genomics is being used to explore the activity of individual genes and to look at the variety of proteins produced by the organism when it faces different environmental challenges. The adaptability of C. jejuni derives from a powerful set of regulatory genes which enable it to change its metabolism rapidly depending on its environment, for example whether it is in contaminated raw chicken, or residing in the human gut. Applying microbial genomic research a food sample can easily be tested for the presence of those genes, which would indicate a contamination with C. jejuni. This test is significantly faster than conventional techniques, and might enable the use of sterilization methods less damaging to the food product.

Further Applications

Application of microbial genomics is not limited to preservation technologies. In principle, all processes in which living (micro-) organisms are involved are amenable to the concept.

Genomics of food microbes generates valuable knowledge that can be used for metabolic engineering, improving cell factories and development of novel preservation methods. Furthermore, pre- and probiotic studies, characterization of stress responses, studies of microbial ecology and, last but not least, development of novel risk assessment procedures will be facilitated. Genomics technology can even be applied as measures for traceability from farm to table.

11

Chemicals: Lead, Mercury, Cadmium and Other Metals

Metals play an important role in human biology, and trace amounts of some metals — manganese, for example — are essential to life. At higher concentrations, however, these same metals are toxic. In addition, some metals — lead, for example — do not occur naturally in the body, and their presence, usually as a result of occupational or pollution-related exposure, is detrimental to health. A number of potentially toxic metals have been reported in breast milk, including lead, mercury and cadmium. Unlike the persistent organic pollutants (POPs), metals do not accumulate in fat, and so do not usually achieve higher concentrations in breast milk than in blood. As a result, infants are likely to be exposed to higher levels before birth than during breastfeeding. Nonetheless, learning about metals in breast milk is important for two reasons: first, as a pathway of exposure, and second, as an indicator of likely prenatal exposures.

Many metals that have been reported in breast milk also contaminate drinking water and so can occur in infant formula at levels even higher than in breast milk. For example, an infant's exposure to cadmium from soy infant formula is about 20 times higher than the levels generally found in breast milk. Estimated cadmium intake from powdered formula has been estimated at six times higher than average levels in breast milk.

Health Effects of Metals

Lead has often been called the leading environmental health threat to children. It is toxic to the developing brain, and at high levels results in numerous poisoning symptoms. In addition, at the low doses common today in many countries, lead has subtle effects on neurological functions, including learning, memory and attention span. Because the infant brain is developing rapidly both before birth and for several years after birth, lead exposures during this critical period are particularly detrimental to the future intellectual potential of children.

Mercury occurs in a number of forms in the environment. The most hazardous for children is methyl mercury, although inorganic mercury is also a potential concern. Methyl mercury, like lead, is toxic to the developing brain. While high-dose exposures can result in a cerebral palsy-like syndrome, low dose exposures may cause subtle deficits in learning and memory. Other metals, such as cadmium, arsenic and manganese, have not been as thoroughly studied in breast milk.

Arsenic is known to cause cancer in humans, and high levels of manganese can cause a syndrome that resembles Parkinson's disease. Cadmium is toxic to the male reproductive system, the kidneys, bone and the brain. All of these contaminants are more likely to affect bottle-fed infants because they are water contaminants and are often found at higher concentrations in infant formula as compared with breast milk.

Human Exposure

People can be exposed to metals in a number of ways, including at work in certain industries, from drinking contaminated water and eating contaminated food, or in hobbies that involve working with metals.

Lead exposure stems primarily from its use in gasoline, paint, water pipes and the lining of food cans. These uses have been banned in many countries, but still persist in many parts of the world. In addition, old, peeling paint and old water pipes can still cause exposures. Other common sources of lead include:

- painting or removing old paint;
- construction work;
- battery manufacturing or recycling;
- automobile repair;
- electronics work;
- ceramics and pottery glazed with lead;
- welding and soldering;
- firearm shooting and cleaning;
- jewellery making and repair;

- stained-glass-window making; and
- cosmetics, including certain hair dyes and kohl.

Most people are exposed to methyl mercury from fish, particularly such predator fish as swordfish, shark and tuna. Freshwater fish from contaminated lakes, rivers and estuaries can also bioaccumulate very high levels of methyl mercury, which are passed on to humans who eat the fish. Other sources of mercury include coal burning, incineration, chlorine manufacturing and mining, as well as some natural sources.

Inorganic mercury exposure primarily comes from dental amalgam fillings. Exposure to cadmium often comes as a result of work or through hobbies, including metal plating, semiconductor manufacture, welding, soldering, ceramics and painting. In addition, it is a contaminant of drinking water, air and food, particularly shellfish. One other important source of cadmium is cigarette smoke; smokers typically have blood levels of cadmium approximately twice those of nonsmokers.

Breast Milk Overview

In general, the metals found in breast milk are usually at lower levels than are found in maternal blood. Thus, breast milk is not the primary pathway of exposure for infants; prenatal trans-placental exposure is a much greater concern. That said, instances of high exposure through breast milk do occur, and are often important indicators of an infant's total exposure. One study found that longer duration of breastfeeding was associated with poorer infant growth in children whose mothers had higher levels of mercury in their bodies. Generally, infants fed formula made with tap water are at the highest risk from metals contaminating the water supply.

Bans and Restrictions

In recent years, some bans and restrictions regarding certain heavy metals have been implemented. For example, lead has been banned from use in gasoline, paint, can linings, or water pipes in more than 50 countries.

The past few years have brought a great deal of state and local legislation restricting the use of mercury in products. For example, several states have passed laws or regulations governing the sale and use of mercury, requiring recycling and imposing notification requirements on the substance and many products containing it. There are many restrictions already in place relating to mercury containing thermometers, and many local communities have set up mercury fever thermometer exchange programs. Several states have adopted bans on offering for sale or use, or distributing for promotional purposes, mercury-added novelties. These are products intended mainly for personal or household enjoyment or for adornment, such as toys, games, ornaments, holiday decorations, apparel, jewellery, figurines and yard statues. Cadmium has also been the subject of legislation. The European Union banned the use of cadmium in materials and components of vehicles put on the market after July 1, 2003, and in new electrical and electronic equipment after July 1, 2006.

Some countries have also had success reducing pollution from incinerators, power plants and factories, thereby reducing emissions of mercury and other metals.

Benchmarks and Exposure Limits

The U.S. EPA has set an action level for lead in water of 15 parts per billion (ppb) and an action level for inorganic mercury of 2 ppb. While the median level of lead in breast milk worldwide is only one-third as high as the U.S. drinking water limit, the most exposed populations have lead levels in breast milk that exceed this limit threefold. Average levels of mercury in breast milk are near the action level for water.

The World Health Organization (WHO) has set a daily permissible intake (DPI) level of 5 micrograms per kilogram per day (μg/kg/day) of lead for children, and the DPI for cadmium is 1 μg/kg/day (for an adult). The U.S. Agency for Toxic Substance and Disease Registry (ATSDR) has established a minimal risk level (MRL) for mercury of 2 μg/kg/day for inorganic mercury and 0.12 μg/kg/day for methyl mercury.

On average, breastfeeding infants are unlikely to exceed these levels. However, in a few polluted communities around the world, infant exposures do exceed these levels.

In the United States, the Health Resources and Services Administration has published a blood lead action level for breastfeeding women of 40 micrograms per deciliter (μg/dL/day)of blood or above.

Women with lead levels this high in their blood may be encouraged to choose alternatives to breastfeeding. Most women have blood lead levels far below 40 micrograms per deciliter. However, recent studies looking at children with blood lead levels of 10 micrograms per dL or less and pregnant women with blood lead levels less than 5 microg/dL have demonstrated significant effects on the children's memory and cognitive functioning. This suggests that the current lead standards may still be too high.

Breast Milk Monitoring Studies

Metals have been detected in breast milk around the world. Countries that have conducted studies detecting one or more of the three major metals of concern (cadmium, lead and mercury) include:

Austria	Hungary	Netherlands	Spain
Belgium	India	New Zealand	Sweden
Bulgaria	Iraq	Nigeria	Thailand
Canada	Italy	Pakistan	Turkey
Finland	Japan	Philippines	United States
Germany	Malaysia	Poland	United Kingdom
Guatemala	Mexico	Romania	Yugoslavia

Levels of lead, manganese and mercury vary widely in breast milk samples around the world, with very high levels detected in some places. Results of a WHO study on trace elements in breast milk are summarized below. These levels reflect both maternal-absorbed dose of these metals and infant exposure, and illustrate the large ranges of exposure across the population. Among women who eat a lot of fish, for example, levels of mercury in breast milk may exceed levels in unexposed women by 100 times.

Table: *Range of Metals Detected in Breast Milk Around the World*

Metal	Median (ppb)	Range (ppb)
Arsenic	0.3	0.1-0.8
Cadmium	0.1	0.1-3.8
Lead	5.0	0.0-41.1
Manganese	18.0	7-102.0
Mercury	2.7	0.64-257.1

Source: World Health Organization, 1993

ppb = parts per billion

Important Case Examples: Several specific examples of metals in the human body merit examination.

Lead

The elevated presence of lead in human blood samples has been an issue for decades, chiefly because of two products: lead-based gasoline and paint. Lead is ubiquitous in the environment as a result, and in many areas of the world, a significant level of lead turns up in breast milk (5 to 20 ppb). That said, lead does not concentrate in breast milk because it does not attach to fat; indeed, levels of lead are generally higher in a mother's blood than in her milk. Several studies have found higher blood lead levels in formula-fed infants than in breast-fed infants. This may be a result of contaminated formula cans or formula prepared using tap water with high lead levels. Lead levels in blood and breast milk correlate closely with areas where lead is still used in gasoline, with the highest levels in areas with heavy traffic. In addition, mothers in countries where lead is still used in gasoline, and mothers living near lead smelters, have higher levels of lead in their breast milk due to community contamination.

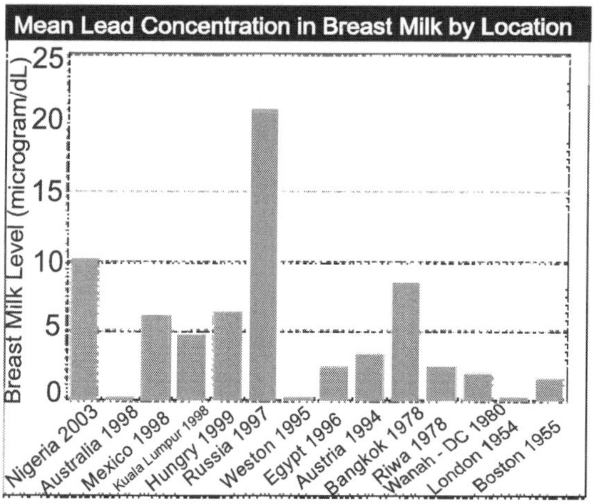

Much of the lead in breast milk does not come from the mothers' exposure during lactation. Instead, it comes from lead stored in the mothers' bones. Because lead mimics the beneficial mineral calcium, it is stored for decades along with calcium in the bones.

During pregnancy and lactation, a woman's body extracts calcium from her own bones to provide calcium for her child's bone development. Calcium extraction from bone is greatest during lactation, and as a result, lead stored in the mother's bones also enters the blood and breast milk during pregnancy and lactation, posing an exposure risk to the fetus. A study in April 2003 confirmed that ensuring adequate dietary calcium intake or taking a calcium supplement before pregnancy, during pregnancy and during the entire lactation period decreases the blood lead level in lactating women.

Supplemental dietary calcium most likely decreases the amount of calcium and lead that comes out of the mother's bones. Therefore, women can significantly reduce their baby's exposure to lead by getting adequate dietary calcium or taking a calcium supplement during pregnancy and lactation.

Mercury

Breast milk levels of mercury are usually lower than levels of lead. Mercury does not accumulate in breast milk; in fact, the levels in the mother's blood are generally about three times higher than the levels in milk. Therefore, prenatal exposure is probably more important than lactation exposure to mercury, in most cases. Two major forms of mercury can enter breast milk. The most hazardous, methyl mercury, does not enter breast milk at high rates because it is attached to red blood cells. But what little does get into breast milk is easily absorbed in the intestine of a nursing infant. The second form, inorganic mercury, enters breast milk easily but is not well absorbed in the infant's gastrointestinal system. One Swedish study found that the mercury in breast milk in the early months of breastfeeding was primarily inorganic mercury from dental amalgam fillings in the mother's mouth. However, after two months of lactation, mercury found in milk was primarily from methyl mercury associated with the mother's fish consumption, rather than dental amalgam fillings. In the past, mercury has been responsible for several mass poisonings — in Minamata, Japan, and in Iraq. In both cases, food contaminated with methyl mercury led to illness and death. Some of those affected were breastfeeding children whose mothers had eaten the contaminated food. However, in both of these scenarios, the levels of mercury were extremely high. The average levels found in women's breast milk today are far lower than in those cases. In one recent study of mercury exposure, breast-fed infants tended to have higher residues of mercury detectable in their hair. The infants with higher hair mercury levels also had improved neurological development, including faster progression to sitting, creeping and standing. Because mercury is known to affect neurological development adversely, the faster development in infants with higher mercury levels was attributed to the benefits of breastfeeding. Thus any possible adverse effects of mercury in breast milk were overcome by the advantages of breastfeeding.

A recent study on laboratory rats lends support to the merits of breastfeeding with respect to mercury. The study looked at the effects of diet that contained a moderate dose of mercury — 5 parts per million — fed to pregnant rats before and during pregnancy and during lactation. The newborn rats on the day of birth actually had 1.5 times more mercury in their blood than their mother. However, during the lactation period the babies' blood mercury levels dropped dramatically. Even though the breast milk contained some methyl-mercury, this study suggests that the exposure of the baby to mercury during breastfeeding is much lower than during fetal development. The diet used in this study simulated that of a high fish consuming-population. After weaning, the newborns were started on a mercury-containing diet and showed an increase in their blood mercury levels. However, a human study in the Faroe Islands, where people eat fish and whale meat, found that those children who were breastfed exclusively for six months ended up shorter and thinner than their peers at 18 months of age. When the researchers looked for an explanation for this finding, they discovered that the children who were exposed to the highest levels of mercury before birth and during breastfeeding were the ones with growth delays.

A number of studies have examined the protective effects of Vitamin E on mercury, finding that vitamin E may decrease the toxic effects of mercury. In addition, the trace element selenium may enhance the ability to clear mercury from the body's cells. This data further supports the role of supplemental multivitamins during pregnancy.

Cadmium

Cadmium levels in breast milk are significantly associated with cigarette smoking. One German study showed a direct relationship between the number of cigarettes a mother smokes per day and the level of cadmium in her breast milk.

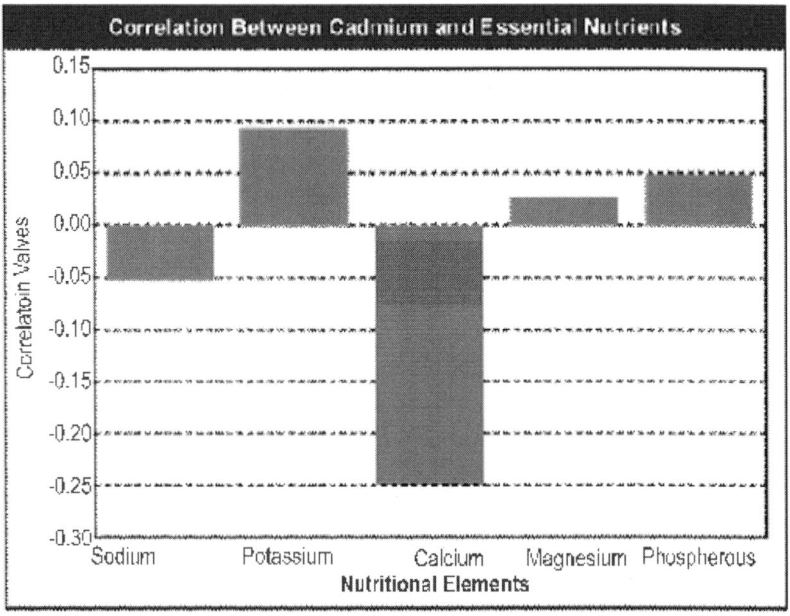

A study in Japan investigated the interaction between some trace metals, including cadmium, and nutritional elements in breast milk. Researchers found a significant association between breast milk cadmium concentration and calcium secretion in breast milk. Increased cadmium in the breast milk appeared to decrease the amount of calcium secreted in the breast milk. In the study, the average concentration of cadmium in breast milk was 0.28 micrograms/liter. Studies have shown that cadmium toxicity targets the kidneys and bone, two crucial calcium metabolizing sites, thus decreasing the amount of calcium in blood and ultimately in breast milk. This data, in conjunction with previously noted data on blood lead levels and calcium, strongly supports previous conclusions about the role of calcium supplementation during pregnancy and lactation, and indicates that cadmium exposure may result in insufficient levels of calcium in the breast milk.

Minerals: Beneficial and Toxic

Vitamins and trace minerals are necessary components of the human diet because they are either inadequately synthesized or not synthesized in the body.

Element	*Breast Milk*	*Formula Concentration Concentration (μg/dL)*
Average (μg/dL)		
Cobalt (Co)	0.2	0.9
Chromium (Cr)	24.3	6.9
Copper (Cu)	400	1207
Iron (Fe)	380	9227
Manganese (Mn)	6.3	46.1
Nickel (Ni)	0.8	14.3
Selenium (Se)	17	Not Quantified
Vanadium (V)	0.2	0.7
Silver (Ag)	0.4	0.7
Aluminium (Al)	67	210.5
Arsenic (As)	6.7	Not Quantified
Titanium (Ti)	6.3	13

However, only small amounts of these elements are needed to carry out the necessary biological reactions. In larger doses, the trace minerals and metals can have toxic effects. In the past, the exact concentrations of many of these trace elements in breast milk has been unclear. One study in Germany determined the concentrations of numerous essential trace elements and toxic metals in human milk and selected infant formulas. Most of the concentrations in infant formulas (except in the case of Chromium) were approximately tenfold higher than in human milk. Therefore, the level of exposure through maternal breast milk is significantly lower than the environmental exposure through infant formula. Thus the data strongly supports the role of breastfeeding in limiting the concentration of exposure to certain potentially toxic metals.

The Health Benefits of Raw Milk

There's little mention in the mainstream media these days, of traditional foods having healing properties. Sure, there's a ton of hype touting unfermented soy products, vegetable oils and supplements as modern saviours, but in reality, these items have risk-to-benefit ratios like many drugs do.

Few people are aware that clean, raw milk from grass-fed cows was actually used as a medicine in the early part of the last century. That's right. Milk straight from the udder, a sort of "stem cell" of foods, was used as medicine to treat, and frequently *cure* some serious chronic diseases. From the time of Hippocrates to until just after World War II, this "white blood" nourished and healed uncounted millions.

Clean raw milk from pastured cows is a complete and properly balanced food. You could live on it exclusively if you had to. Indeed, published accounts exist of people who have done just that. What's in it that makes it so great? Let's look at the ingredients to see what makes it such a powerful food.

Proteins

Our bodies use amino acids as building blocks for protein. Depending on who you ask, we need 20-22 of them for this task. Eight of them are considered *essential*, in that we have to get them from our food. The remaining 12-14 we can make from the first eight in the chemical factories of our bodies.

Raw cow's milk has all 8 essential amino acids, saving our bodies the work of having to convert any into usable form. About 80% of the proteins in milk are caseins-reasonably heat stable but easy to digest. The remaining 20% or so fall into the class of whey proteins, many of which have important physiological effects (bioactivity). Also easy to digest, but very heat-sensitive, these include key enzymes (specialized proteins) and enzyme inhibitors, immunoglobulins (antibodies), metal-binding proteins, vitamin binding proteins and several growth factors. Current research is now focusing on fragments of protein (peptide segments) hidden in casein molecules that exhibit anti-microbial activity.

Lactoferrin, an iron-binding protein, has numerous beneficial properties including (as you might guess) improved absorption and assimilation of iron, anti-cancer properties and anti-microbial action against several species of bacteria responsible for dental cavities. Recent studies also reveal that it has powerful antiviral properties as well.

Two other players in raw milk's antibiotic protein/enzyme arsenal are *lysozyme* and *lactoperoxidase*. Lysozyme can actually break apart cell walls of certain undesirable bacteria, while lactoperoxidase teams up with other substances to help knock out unwanted microbes too.

The immunoglobulins, an extremely complex class of milk proteins also known as *antibodies*, provide resistance to many viruses, bacteria and bacterial toxins and may help reduce the severity of asthma symptoms. Studies have shown significant loss of these important disease fighters when milk is heated to normal processing temperatures.

Carbohydrates

Lactose, or milk sugar, is the primary carbohydrate in cow's milk. Made from one molecule each of the simple sugars glucose and galactose, it's known as a disaccharide. People with lactose intolerance for one reason or another (age, genetics, etc.), no longer make the enzyme lactase and so can't digest milk sugar. This leads to some unsavory symptoms, which, needless to say, the victims find rather unpleasant at best. Raw milk, with its lactose-digesting Lactobacilli bacteria intact, may allow people who traditionally have avoided milk to give it another try.

The end-result of lactose digestion is a substance called lactic acid (responsible for the sour taste in fermented dairy products). Besides having known inhibitory effects on harmful species of bacteria, lactic acid boosts the absorption of calcium, phosphorus and iron, and has been shown to make milk proteins more digestible by knocking them out of solution as fine curd particles.

Fats

Approximately two thirds of the fat in milk is saturated. Good or bad for you? Saturated fats play a number of key roles in our bodies: from construction of cell membranes and key hormones to providing energy storage and padding for delicate organs, to serving as a vehicle for important fat-soluble vitamins.

All fats cause our stomach lining to secrete a hormone (cholecystokinin or CCK) which, aside from boosting production and secretion of digestive enzymes, let's us know we've eaten enough. With that trigger removed, non-fat dairy products and other fat-free foods can potentially help contribute to over-eating. Consider that, for thousands of years before the introduction of the hydrogenation process (pumping hydrogen gas through oils to make them solids) and the use of canola oil (from genetically modified rapeseed), corn, cottonseed, safflower and soy oils, dietary fats were somewhat more often saturated and frequently animal-based. (Prior to about 1850, animals in the U.S. were not so heavily fed corn or grain). Use of butter, lard, tallows, poultry fats, fish oils, tropical oils such as coconut and palm, and cold pressed olive oil were also higher than levels seen today.

Now consider that prior to 1900, very few people died from heart disease. The introduction of hydrogenated cottonseed oil in 1911 (as trans-fat laden Crisco) helped begin the move away from healthy animal fats, and toward the slow, downward trend in cardiovascular health from which millions continue to suffer today.

CLA, short for *conjugated linoleic acid* and abundant in milk from grass-fed cows, is a heavily studied, polyunsaturated Omega-6 fatty acid with promising health benefits. It certainly does wonders for rodents, judging by the hundreds of journal articles I've come across! There's serious money behind CLA, so it's a sure bet there's something to it.

Among CLA's many potential benefits: it raises metabolic rate, helps remove abdominal fat, boosts muscle growth, reduces resistance to insulin, strengthens the immune system and lowers food allergy reactions. As luck would have it, grass-fed raw milk has from 3-5 times the amount found in the milk from feed lot cows.

Vitamins

Volumes have been written about the two groups of vitamins, water and fat soluble, and their contribution to health. Whole raw milk has them all, and they're completely available for your body to use. Whether regulating your metabolism or helping the biochemical reactions that free energy from the food you eat, they're all present and ready to go to work for you.

Just to repeat, nothing needs to be added to raw milk, especially that from grass-fed cows, to make it whole or better. No vitamins. No minerals. No enriching. It's a complete food.

Minerals

Our bodies, each with a biochemistry as unique as our fingerprints, are incredibly complex, so discussions of minerals, or any nutrients for that matter, must deal with ranges rather than specific amounts. Raw milk contains a broad selection of completely available minerals ranging from the familiar calcium and phosphorus on down to trace elements, the function of some, as yet, still rather unclear.

A sampling of the health benefits of calcium, an important element abundant in raw milk includes: reduction in cancers, particularly of the colon: higher bone mineral density in people of every age, lower risk of osteoporosis and fractures in older adults; lowered risk of kidney stones; formation of strong teeth and reduction of dental cavities, to name a few.

An interesting feature of minerals as nutrients is the delicate balance they require with other minerals to function properly. For instance, calcium needs a proper ratio of two other macronutrients, phosphorus and magnesium, to be properly utilized by our bodies. Guess what? Nature codes for the entire array of minerals in raw milk (from cows on properly maintained pasture) to be in proper balance to one another thus optimizing their benefit to us.

Enzymes

The 60 plus (known) fully intact and functional enzymes in raw milk have an amazing array of tasks to perform, each one of them essential in facilitating one key reaction or another. Some of them are native to milk, and others come from beneficial bacteria growing in the milk. Just keeping track of them would require a post-doctoral degree!

To me, the most significant health benefit derived from food enzymes is the burden they take off our body. When we eat a food that contains enzymes devoted to its own digestion, it's that much less work for our pancreas. Given the choice, I'll bet that busy organ would rather occupy itself with making metabolic enzymes and insulin, letting food digest itself.

The amylase, bacterially-produced lactase, lipases and phosphatases in raw milk, break down starch, lactose (milk sugar), fat (triglycerides) and phosphate compounds respectively, making milk more digestible and freeing up key minerals. Other enzymes, like catalase, lysozyme and lactoperoxidase help to protect milk from unwanted bacterial infection, making it safer for us to drink.

Cholesterol

Milk contains about 3mg of cholesterol per gram-a decent amount. Our bodies make most of what we need, that amount fluctuating by what we get from our food. Eat more, make less. Either way, we need it. Why not let raw milk be one source?

Cholesterol is a protective/repair substance. A waxy plant steroid (often lumped in with the fats), our body uses it as a form of water-proofing, and as a building block for a number of key hormones.

It's natural, normal and essential to find it in our brain, liver, nerves, blood, bile, indeed, every cell membrane. The best analogy I've heard regarding cholesterol's supposed causative effects on the clogging of our arteries is that blaming it is like blaming crime on the police because they're always at the scene.

Seriously consider educating yourself fully on this critical food issue. It could, quite literally, save your life.

Beneficial Bacteria

Through the process of fermentation, several strains of bacteria naturally present or added later (*Lactobacillus, Leuconostoc* and *Pediococcus,* to name a few) can transform milk into an even more digestible food.

With high levels of lactic acid, numerous enzymes and increased vitamin content, 'soured' or fermented dairy products like yogurt and kefir (made with bacteria and *yeast*, actually) provide a plethora of health benefits for the savvy people who eat them. Being acid lovers, these helpful little critters make it safely through the stomach's acid environment to reach the intestines where they really begin to work their magic

Down there in the pitch black, some of them make enzymes that help break proteins apart-a real benefit for people with weakened digestion whether it be from age, pharmaceutical side-effects or illness.

Other strains get to work on fats by making lipases that chop triglycerides into useable chunks. Still others take on the milk sugar, lactose, and, using fancy sounding enzymes like beta-galactosidase, glycolase and lactic dehydrogenase (take notes, there'll be a quiz later!), make lactic acid out of it.

As I mentioned way up yonder in the Carbohydrate section, having lactic acid working for you in your nether regions can be a good thing. Remember? It boosts absorption of calcium, iron and phosphorus, breaks up casein into smaller chunks and helps eliminate bad bugs.

Raw milk is a living food with remarkable self-protective properties, but here's the kick: most foods tend to go south as they age, raw milk just keeps getting better.

Not to keep harping on this, but what the heck: through helpful bacterial fermentation, you can expect an increase in enzymes, vitamins, mineral availability and overall digestibility. Not bad for old age!

A Word About Diet In General

Use common sense and stick with whole, unprocessed foods, free from genetic tweaking (there's still just too much conflicting information out there on that topic), and you'll likely be ahead of the game.

Cook your foods minimally, and you'll be even better off. Learn about sprouting and fermentation. Question everything before letting it past your lips.

Explore what worked for countless generations before ours, and put it to work for yourself today. You *can* achieve great health by diet alone. I've done it, and so can you!

Few people are aware that clean, raw milk from grass-fed cows was actually used as a medicine in the early part of the last century. From the time of Hippocrates to until just after World War II, this "white blood" nourished and healed millions.

Clean raw milk from pastured cows is a complete and properly balanced food which you could live on exclusively if you had to.

What's in it that makes it so great? Let's look at the ingredients to see what makes it such a powerful food.

Proteins: Our bodies use amino acids as building blocks for protein. Depending on who you ask, we need 20-22 of them for this task. Eight are considered essential, in that we have to get them from our food. The remaining 12-14 we can make from the first eight in the chemical factories of our bodies.

Raw cow's milk has all 20 of the standard amino acids. About 80% of the proteins in milk are caseins; reasonably heat stable but easy to digest. The remaining 20% fall into the class of whey proteins, many of which have important physiological effects (bioactivity). Also easy to digest, but very heat sensitive, these include key enzymes (specialized proteins) and enzyme inhibitors, immunoglobulins (antibodies), metal-binding proteins, vitamin binding proteins and several growth factors.

Lactoferrin, an iron-binding protein, has numerous beneficial properties including improved absorption and assimilation of iron, anti-cancer properties and anti-microbial action against several species of bacteria responsible for dental cavities. Recent studies also reveal that it has powerful antiviral properties as well.

Two other players in raw milk's antibiotic protein/enzyme arsenal are lysozyme and lactoperoxidase. Lysozyme can actually break apart cell walls of certain undesirable bacteria, while lactoperoxidase teams up with other substances to help knock out unwanted microbes.

The immunoglobulins, an extremely complex class of milk proteins known as antibodies, provide resistance to many viruses, bacteria and bacterial toxins and may help reduce the severity of asthma symptoms. Studies have shown significant loss of these important disease fighters when milk is heated to normal processing temperatures.

Carbohydrates: Lactose, or milk sugar, is the primary carbohydrate in cow's milk. Made from one molecule each of the simple sugars glucose and galactose, it's known as a disaccharide. People with lactose intolerance for one reason or another (age, genetics, etc.), no longer make the enzyme lactase and so can't digest milk sugar. This leads to some unsavory symptoms, which, needless to say, the victims find rather unpleasant at best. Raw milk, unlike pasteurized, has it's milk sugar enzyme, lactase, undamaged, and so, may allow people who traditionally have avoided milk to give it another try.

Lactose provides milk's natural complement of Lactobacilli bacteria with the raw material from which to make lactic acid (the sour taste in fermented dairy products). Besides having known inhibitory effects on harmful species of bacteria, lactic acid boosts the absorption of calcium, phosphorus and iron, and has been shown to make milk proteins more digestible.

Fats: Approximately two thirds of the fat in milk is saturated. Saturated fats play a number of key roles in our bodies: from construction of cell walls and key hormones to providing energy storage and padding for delicate organs, to providing a vehicle for carrying important fat-soluble vitamins

All fats cause our stomach lining to secrete a hormone (CCK) which, aside from boosting production and secretion of digestive enzymes, let's us know we've eaten enough. With that trigger removed, non-fat dairy products and other fat-free foods can potentially help contribute to over-eating.

Consider that, for thousands of years before the introduction of the hydrogenation process (pumping hydrogen gas through oils to make them solids) and the use of canola oil (from genetically modified rapeseed), corn, cottonseed, safflower and soy oils, dietary fats were largely saturated and often animal based. Healthy cultures all over the world thrived on the use of butter, lard, tallows, poultry fats, fish oils, tropical oils such as coconut and palm, and cold pressed olive oil.

Now consider that prior to 1900, very few people died from heart disease. The introduction of hydrogenated cottonseed oil in 1911 helped begin the move away from healthy animal fats, and toward the slow, downward trend in cardiovascular health from which millions continue to suffer today.

CLA, short for conjugated linoleic acid and abundant in milk from grass-fed cows, is a polyunsaturated Omega-6 fatty acid with promising health benefits. CLA raises metabolic rate, helps remove abdominal fat, boosts muscle growth, reduces resistance to insulin, strengthens the immune system and lowers food allergy reactions. Grass-fed raw milk has 3-5 times the amount found in the milk from feed lot cows.

Vitamins: Volumes have been written about the two groups of vitamins, water and fat soluble, and their contribution to health. Whole raw milk has them all. Whether regulating your metabolism or helping the biochemical reactions that free energy from the food you eat, they're all present and ready to go to work for you.

Nothing needs to be added to raw milk to make it whole or better. No vitamins. No minerals. No enriching. It's a complete food. Heated milk must have destroyed components added back in—especially the important fat soluble vitamins A and D.

Minerals: Raw milk contains a broad selection of completely available minerals ranging from calcium and phosphorus down to trace elements.

A sampling of the health benefits of calcium, a 'macronutrient' abundant in raw milk includes: reduction in cancers, particularly of the colon; higher bone mineral density in people of every age, lower risk of osteoporosis and fractures in older adults; lowered risk of kidney stones; formation of strong teeth and reduction of dental cavities, to name a few.

An interesting feature of minerals as nutrients is the delicate balance they require with other minerals to function properly. For instance, calcium needs a proper ratio of two other macronutrients, phosphorus and magnesium, to be properly utilized by our bodies. Guess what? The entire array of minerals in raw milk is in proper balance to one another thus optimizing their benefits.

Enzymes: The most significant health benefit derived from food enzymes is the burden they take off our body. When we eat a food that contains enzymes devoted to its own digestion, it's that much less work for our pancreas.

The amylase, lactase, lipase and phosphatase in raw milk, break down starch, lactose (milk sugar), fat (triglycerides) and phosphate compounds respectively, making milk more digestible and freeing up key minerals. Other enzymes, like catalase, lysozyme and lactoperoxidase help protect milk from unwanted bacterial infection.

Cholesterol: Milk contains about 3mg of cholesterol per gram—a decent amount. Our bodies make most of what we need, that amount fluctuating by what we get from our food. Cholesterol is a protective/repair substance. A waxy plant steroid (often lumped in with the fats), our body uses it as a form of water-proofing, and as a building block for key hormones.

It's natural, normal and essential to find it in our brain, liver, nerves, blood, bile, indeed, every cell wall. Seriously consider educating yourself fully on this critical food issue. It could, quite literally, save your life.

Beneficial Bacteria: Through the process of fermentation, several strains of bacteria naturally present or added later (Lactobacillus, Leuconostoc and Pediococcus, to name a few) can transform milk into an even more digestible food. With high levels of lactic acid, numerous enzymes and increased vitamin content, 'soured' or fermented dairy products like yogurt and kefir (made with bacteria and yeast) provide a plethora of health benefits. Being acid lovers, these helpful little critters make it safely through the stomach's acid environment to reach the intestines where they begin to work their magic.

Some of them make enzymes that help break proteins apart—a real benefit for people with weakened digestion whether it be from age, pharmaceutical side-effects or illness. Other strains get to work on fats by making lipases that chop triglycerides into useable chunks. Still others take on the milk sugar, lactose, and, using enzymes like beta-galactosidase, glycolase and lactic dehydrogenase make lactic acid out of it.

Raw milk is a living food with remarkable self-protective properties. Perhaps it is time to explore all the benefits of raw milk and begin living a healthy lifestyle. Explore what worked for countless generations before ours, and put it to work for yourself today.

Milk and Milk Products

Although cow's milk is the most popular in many countries, milk can be obtained from many different sources.

For example, milk from goats and sheep makes a substantial contribution to the total milk production in countries of Eastern and Southern Europe, Malawi, and Barbados, whereas the water buffalo is a common source of milk in much of Asia.

Woman Milking a Goat

Milk is a perishable commodity and spoils very easily. Its low acidity and high nutrient content make it the perfect breeding ground for bacteria, including those which cause food poisoning (pathogens).

Bacteria from the animal, utensils, hands, and insects may contaminate the milk, and their destruction is the main reason for processing. This preservation of the milk can be achieved by fermentation, heating, cooling, removal of water, and by concentration or separation of components, to produce foods such as butter or cheese.

The degree to which milk consumption and processing occurs will differ from region to region. It is dependent upon a whole host of factors, including geographic and climatic conditions, availability and cost of milk, food taboos, and religious restrictions. Where processing does exist, many traditional techniques can be found for producing indigenous milk products. These are more stable than raw milk and provide a means of preservation as well as adding variety to the diet. In addition, the introduction of western-style dairy products and the subsequent setting up of small-scale dairies has provided more choice of dairy products to the consumer.

Nutritional Significance

Milk is often regarded as being nature's most complete food. It earns this reputation by providing many of the nutrients which are essential for the growth of the human body. Being an excellent source of protein and having an abundance of vitamins and minerals, particularly calcium, milk can make a positive contribution to the health of a nation. The realization of its nutritional attributes is clearly illustrated by the implementation of numerous 'school milk programmes' worldwide.

Fermented-milk products such as yoghurt and soured milk contain bacteria from the *Lactobacilli* group. These bacteria occur naturally in the digestive tract and have a cleansing and healing effect. Therefore the introduction of fermented products into the diet can help prevent certain yeasts and bacteria which may cause illness.

Many people suffer from a condition known as 'lactose intolerance'. This means that they are unable to digest the milk fat (lactose). Such people can, however, tolerate milk if it is fermented to produce foods such as yoghurt. During fermentation, lactic acid producing bacteria break down lactose, and in doing so eliminate the cause of irritation.

The Quality of Milk

The type of animal, its quality, and its diet can lead to differences in the colour, flavour, and composition of milk. Infections in the animal which cause illness may be passed directly to the consumer through milk. It is therefore extremely important that quality-control tests are carried out to ensure that the bacterial activity in raw milk is of an acceptable level, and that no harmful bacteria remain in the processed products.

Table: *Average composition (%) of milks of various mammals*

Species	*Water*	*Fat*	*Protein*	*Lactose*	*Ash*
Human	87.43	3.75	1.63	6.98	0.21
Cow	87.2	3.7	3.5	4.9	0.7
Goat	87.00	4.25	3.52	4.27	0.86
Sheep	80.71	7.9	5.23	4.81	0.9
Indian buffalo	82.76	7.38	3.6	5.48	0.78
Camel	87.61	5.38	2.98	3.26	0.7
Horse	89.04	1.59	2.69	6.14	0.51
Llama	86.55	3.15	3.9	5.6	0.8

Standard Testing Procedures

Milk Fat

The price paid for milk is usually dependent upon the milk-fat content, and this may be determined either at the collection stage or at the dairy using a piece of equipment known as a butyrometer. Additionally the specific gravity can be measured using a hydrometer. This can also be used as an aid to detect adulteration.

Bacterial Activity

Routinely it is necessary to check the microbiological quality of raw milk using either methylene blue or resazurin dyes. These tests indicate the activity of bacteria in the milk sample and the results determine whether the milk is accepted or rejected.

Both tests work on the principle of the time taken to change the colour of the dye. The length of time taken is proportional to the number of microorganisms present (the shorter the time taken, the higher the bacterial activity). It is preferable to use the resazurin test as this is less time-consuming. For these tests, basic laboratory equipment will be needed such as test-tubes, a water bath, accurate measuring equipment, and a supply of dyes.

After collection the milk should ideally be stored at a temperature of 4°C or below. This is necessary to slow the growth of any contaminating bacteria.

Phosphatase Test

For pasteurized milk, it is possible to ensure that pasteurization has been adequately achieved by testing for the presence of the enzyme phosphatase. The destruction of phosphatase is regarded as a reliable test to show that the milk has been sufficiently heat-processed, because this enzyme (present in raw milk) is destroyed by pasteurization conditions.

It is stressed that pasteurization is an effective safeguard against spoilage and food poisoning only if the milk is not re-contaminated after pasteurization.

Processing

Table: *Equipment required*

Processing stage	*Equipment*	*Section reference*
Store at 4°C	Refrigerated storage	15.0
	Thermometer	63.0
Test for fat content	Butyrometer	64.5
Test specific gravity	Hydrometer	64.4
Test bacterial activity	Supply of dyes	64.6
	ThermometerBasic laboratory equipment is required for most of the tests	63.0
Filter	Filter cloth	0.80
	Filter press	29.2
Homogenization	Homogenizer	37.0
Fill into bottles	Liquid-filling machine	28.1
	Capping machine	47.2

Pasteurization	Boiling pan	48.0
	or pasteurizer	50.0
	Heat source	36.0
	Thermometer	63.0
Sterilization	Pressure cooker	48.0
	Retort	05.1
	Heat source	36.0
	Thermometer	63.0
Cool	Bottle-cooling system	

Liquid milk

Milk can be kept for longer periods of time if it is heated to destroy the bacteria or cooled to slow their growth. Pasteurization and sterilization are the two most commonly-used heat treatments. Technically, it is possible for both to be carried out on a small scale, but they are most usually performed on a larger industrial scale due to the need for qualified, experienced staff and accurate and strictly controlled hygienic processing conditions.

Homogenization : Homogenization breaks up the oil droplets in milk and prevents the cream from separating out and forming a layer. This is of particular importance for sterilized milk which has a long shelf-life and when the formation of a cream layer is not desired. Additional changes include increased viscosity and a richer taste. Homogenizers are more usually designed for industrial-scale production, but it is possible to purchase smaller versions.

Filling : The most common packaging material for both pasteurized and sterilized milk is glass bottles sealed with either foil or metal caps, although plastic bottles, plastic bags, and cardboard cartons are all used when bottles are not available or too expensive.

Pasteurization

Pasteurization is a relatively mild heat treatment, (usually performed below 100°C) which is used to extend the shelf-life of milk for several days. It preserves the milk by the inactivation of enzymes and destruction of heat-sensitive microorganisms, but causes minimal changes to the nutritive value or sensory characteristics of a food.

Some heat-resistant bacteria survive to spoil the milk after a few days, but these bacteria do not cause food poisoning. The time and temperature combination needed to destroy 'target' microorganisms will vary according to a number of complex inter-related factors. For milk, the heating time and temperature is either 63°C for 30 minutes or alternatively 72°C for 15 seconds.

Only the former combination is possible on a small scale and for this the simplest equipment required is an open boiling pan. Better control is achieved using a steam jacketed pan, and this can be fitted with a stirrer to improve the efficiency of heating. Both of these are batch processes which are suited to small-scale operation. A higher production rate may be possible using a tubular-coil pasteurizer. This equipment has been tested and has been successful for some fruit products but it is presently still at a developmental stage.

Sterilization

Sterilization is a more severe heat treatment designed to destroy all contaminating bacteria. The milk is sterilized at a temperature of 121°C maintained for 15-20 minutes. This can be achieved using a retort or pressure cooker. Unlike pasteurization, this process causes substantial changes to the nutritional and sensory quality of the milk. In some countries, flavoured milk has become a very popular product.

However, sterilization is not recommended for small-scale production for the following reasons:

- The cost of a retort and ancillary equipment is high for the small-scale processor.
- It is essential that the correct heating conditions are carefully established and maintained for every batch of milk that is processed. If the milk is overheated, the quality is reduced, and it may have a rather burnt taste and aroma.
- If the milk is not heated sufficiently, there is a risk that microorganisms will survive and grow inside the bottle. In low-acid foods such as milk, many types of bacteria including *Clostridium botulinum* can grow and cause severe food poisoning.
- Due to the potential dangers from food poisoning, the skills of a qualified food technologist/microbiologist are required in order to routinely examine samples of sterilized milk that have been subjected to accelerated storage conditions. This requires a supply of microbiological media and equipment.

In summary, the process of sterilization requires a considerable capital investment, the need for trained and experienced staff, regular maintenance of sophisticated equipment, and a comparatively high operating expenditure.

Cooling

Pasteurization does not destroy all of the microorganisms, therefore the milk has to be cooled rapidly to prevent the growth of surviving bacteria. Cooling can be achieved on a small scale by using a bottle-cooling system.

Storage

Pasteurized milk has a shelf-life of 2-3 days if kept at 4°C. Maintaining this low temperature causes a substantial increase to the cost of transportation and distribution and is therefore a major disadvantage to the development of a small-scale pasteurized milk business. If packaged in sealed bottles and stored at room temperature, sterilized milk should have a shelf-life in excess of six months.

Churning Cream

Churning is continued until fat granules are present and at this stage the mixture is drained to remove liquid that has separated from the granules. This liquid is known as buttermilk and can be used as either a beverage or as an ingredient in animal feed.

Washing

Clean water equivalent in weight to the buttermilk is added to the churn in order to wash the butter granules. The wash water is drained away. Churning is continued for a short time to compact the butter, and once this has been achieved it is removed from the churn.

Forming and Packaging

Butter is kneaded to achieve a smooth and pliable texture. This can be done using simple hand-tools such as butter pats. Alternatively for higher production rates a specially-designed kneader can be used. Once the butter has a uniform and smooth texture it is formed into blocks with butter pats and packed in either greaseproof paper or foil wrappers.

Working Butter with Butter Pats

Storage

Due to its high fat composition, butter must be stored at temperatures below 10°C otherwise the fat becomes rancid and imparts undesirable 'off' flavours. The water droplets in butter (20 per cent) can also allow bacteria to grow if it is not kept under cool conditions.

Ghee

Ghee is made from butter which has been heated and clarified. At ambient temperatures it is a semi-solid mass with a granular texture, but on melting (40°C+) it turns into a clear, thin liquid. It has a high demand in some countries for domestic use, as an ingredient for local food production (for example bakeries and confectionery manufacturers), and as an export commodity.

Alternatively, cream is boiled gently to evaporate the water. During boiling the product is stirred continuously until the milk proteins start to coagulate, forming particles, and the colour of the cream darkens. Heating is stopped and the product is left to set. The particles settle at the bottom of the vessel and the milk-fat is separated. The principles of preservation are:

- heating to destroy enzymes and contaminating microorganisms
- to reduce the water-content by evaporation, and in doing so prevent the growth of microorganisms.

Table: *Production stages for ghee*

Ingredients	***Process***	***Equipment***	***Section reference***
Butter	Heating	Heat source	36.0
		Large boiling pan or steam jacketed pan	48.0
	Cool to room temperature	Thermometer	63.0
	Filter	Filter cloth	08.0
	Fill into jars/pots	Funnel or liquid-filling machine	28.1
		Capping machine	47.2
	Store at ambient temperatures		

Packaging and Storage

Metal containers are normally used. They should be thoroughly cleaned, especially if they are re-usable, and they should be made airtight. Alternatives to metal cans include coloured glass jars with metal lids, or ceramic pots sealed with cork/plastic stoppers.

Ghee is usually stored at room temperatures as cold storage affects the granular texture. Thus ghee is useful for those consumers with no access to refrigeration.

Cultured/Fermented Dairy Products

The technology of cultured milk products such as yoghurt, curd, and cheese is based upon the microbial conversion of the milk-sugar lactose to lactic acid (lactic acid accounts for the characteristic 'sourness' of such products). In order for the conversion to take place, lactic acid producing bacteria must be present.

This may occur by allowing the milk to sour naturally, but it is better to introduce the appropriate bacteria as a starter culture. Starter cultures may be in the form of a small quantity of previously-cultured product or may be purchased as a commercially-prepared culture.

Yoghurt/Curd

Yoghurt is a fermented milk product that evolved by allowing naturally-contaminated milk to sour at a warm temperature. Yoghurt can be either unsweetened or sweetened, set, or stirred. Curd is the name given to a yoghurt-type product made from buffalo milk.

The principles of preservation for yoghurt are:

- Pasteurization of the raw milk to destroy contaminating microorganisms and enzymes.
- An increase in acidity due to the production of lactic acid from lactose. This inhibits the growth of food-poisoning bacteria.
- Storage at a low temperature to inhibit the growth of microorganisms.

***Table:** Production stages for set yoghurt*

Ingredients	*Process*	*Equipment*	*Section reference*
Milk and starter culture (2 per cent)	Preheat to 70°C for 15-20 minutes	Heat source	36.0
		Thermometer	63.0
		Boiling pan	48.0
	Cool to 30-40°C	Thermometer	63.0
	Addition of starter culture	Measuring and weighing equipment	64.1 & 64.2
	Pour into bottles/pots	Funnel or Liquid filler	28.1
		Sealing machine	47.1
		or Capping machine	47.2
	Incubate at 43-45°C	Commercial incubator	39.0
		Thermometer	63.0
	Store at 4°C	Refrigerated storage	15.0

Heating

In the manufacture of yoghurt, milk is normally heated to 70°C for 15-20 minutes, using an open boiling pan, or alternatively a steam jacketed pan.

Addition of Starter Culture

The milk is cooled to between 30 40°C and inoculated with a mixed culture of *Lactobacillus bulgaricus* and *Streptococcus thermophilus* (usually in a ratio of 1:1). If a commercial starter-culture is used, the directions for use will be given. However, if a culture from a previous batch is used, then it is usual to add 2-3 tablespoons per litre of prepared milk. Yoghurt of the stirred variety can be fermented in the mixing container. To make set yoghurt the inoculated milk should be poured into the individual pots before fermentation.

Selling Curd from a Roadside Stall

Incubation

The microorganisms that produce yoghurt are most active within a temperature range of 32-47°C. Ambient temperatures are therefore not adequate and a heated incubator is needed. Small commercially-available yoghurt-makers consist of an electrically-heated base and a set of plastic or glass containers. Most yoghurt-makers make four or five individual half litre cups at a time. There are other simple and inexpensive ways of incubating yoghurt such as an insulated box, keeping the jars/pots surrounded by warm water, or by using thermos flasks (the latter is only suitable for stirred yoghurt). Incubation takes approximately five hours. When fermentation is complete, stirred yoghurt is cooled and flavoured or sweetened prior to packaging. In set yoghurt all ingredients are added before fermentation.

Packaging and Storage

Yoghurt or curd is commonly packaged in plastic pots fitted with a plastic lid, or heat-sealed with foil, although traditionally, curd is packaged in clay pots. Such pots are made from local materials and can be re-used or later used for cooking. The shelf-life of yoghurt is usually 3-8 days when stored at temperatures below 10°C.

Cheese

Cheese is made from milk by the combined action of lactic acid bacteria and the enzyme rennin (known as rennet). Just as cream is a concentrated form of milk fat, cheese is a concentrated form of milk-protein. The differences in cheeses that are produced in different regions result from variations in the composition and type of milk, variations in the process, and the bacteria used. The different cheese varieties can be classified as either hard or soft.

Collecting milk for Cheese-making

Hard cheeses such as Cheddar and Edam have most of the whey drained out and are pressed. Soft cheeses such as paneer contain some of the whey and are not pressed. Many indigenous cheeses are soft types. The hardness, flavour, and other qualities of a cheese can be varied by changes to the process conditions, to suit local tastes. However the principal steps of a cheese-making process are basically the same. The principles of preservation are:

- the raw milk is pasteurized to destroy most enzymes and contaminating bacteria

- fermentation by lactic-acid bacteria increases the acidity which inhibits the growth of food-poisoning and spoilage bacteria
- the moisture content is lowered and salt is added to inhibit bacterial and mould growth.

The table, right, outlines the stages of production and the equipment needed to produce Edam cheese.

Pre-heating : The pasteurized milk is heated to a temperature at which the starter-culture can work.

Addition of Starter Culture : Starter-culture is added to the milk at the rate of approximately 2 per cent of the weight of milk. The vessel used should be either aluminium or stainless steel.

Addition of Rennet : The rennet should be 1 per cent of the weight of milk. The rennet alters the milk proteins and allows them to form the characteristic curd.

Incubation: The milk is allowed to stand until it sets to a firm curd.

Treatment of the Curd

The curd is cut into cubes which facilitate the elimination of whey from the gel. The curd is then cooked at 40°C for a period of twenty minutes which has the action of firming the curd. After cooling, the whey is drained off. The curd is pressed to ensure that most of the whey has been removed, and is then cut to fit the cheese-moulds, and finally pressed with weights.

Ripening

This is the final stage in the cheese-making process. It is a process which allows the development of gas in some cheeses and the development of flavour. The longer the ripening process the stronger the flavour. Ripening usually takes place in ripening rooms, where the temperature and humidity must be controlled for the optimum development of the cheese.

Selling ice Cream from a Bicycle

Ice Cream

Ice cream is a frozen mixture which contains milk, sugar, fat, and optional thickeners (e.g. pectin or gelatin), colouring, and flavouring. It may be sweetened and flavoured in numerous ways with nuts, fruit pieces, and natural or artificial flavours and colours. The principles of preservation are:

- pasteurization to destroy most microorganisms and enzymes
- freezing to inhibit microbial growth.

Pasteurization : Pasteurization is carried out by heating to 65°C for a period of 30 minutes.

Cooling and beating : Ice cream is a complex mix of small ice crystals and air bubbles in a milk-fat/water emulsion. To achieve this, it is necessary to cool the mixture quickly to produce small ice crystals and at the same time incorporate air into the product by beating. Ice cream makers are available commercially and work on the following principle. The mixture is placed in a bowl which is kept at a low temperature (either surrounded by ice and salt, or having been chilled in a freezer). It is then agitated by a manually-operated rotor or by a powered stirrer. At the end of this process the ice cream should be at a temperature of approximately-5°C, and be partly frozen.

Packaging and Storage

The ice cream is usually packaged in plastic, waxed paper, or cardboard containers, and is stored at below -18°C. The storage temperature is important for two reasons:

- to maintain the texture of the product
- to prevent the growth of microorganisms.

Ice cream may be transported in an insulated box (e.g. for sale from a bicycle). It is especially important to guard against thawing and re-freezing as this will cause changes in texture and mouthfeel, and there is the increased possibility of food poisoning by contaminating food poisoning microorganisms.

Suitability for Small-scale Production

In some regions, there is a high demand for dairy products both for traditional and modern items. Much of the technology and machinery necessary for processing is fairly simple, which may at first sight appeal to the would-be processor. Milk, however, is a highly perishable food and there is a high risk of transmitting food-poisoning bacteria to consumers. It is stressed, therefore, that milk processing of any kind must be done under carefully controlled hygienic conditions. Furthermore, in developing countries, milk processing is often more problematic than in temperate climates, owing to higher temperatures and humidity. Consequently, milk spoils at a faster rate, cheese ripens too quickly, and it is often difficult to ensure adequate cooling conditions.

Bibliography

Adams, H.P. and R.B. Bushnell.: *Dry Cow Feeding and Management*, A Western Regional Extension Publication, 1979.

Ashraf, Syed Mahboob: *Handbook of Fruit and Vegetable Products*, Agrobios, Delhi, 2008.

Balch C.N.C, P.: *Prescription for Nutritional Healing,* New York, U.S.A: Avery Publishing Group, 1997.

Bamberg, J. H.: *British Petroleum and global oil, 1950-1975: the challenge of nationalism*, Cambridge University Press. 2000.

Basavaraj S. Benni, Rawat: *Dairy Co-operative Management and Practice*, Delhi, 2005.

Bushnell, R.B.: *Dry Cow Feeding and Management*, A Western Regional Extension Publication, 1979.

Cristobal Noe Aguilar*: Food Science and Food Biotechnology in Developing Countries*, Asiatech Pub, 2008.

Devraj B: *Impact of Dairy On Small and Marginal Farmers*, Prateeksha Publications, Delhi, 2010.

Fiechter, Armin: *History of Modern Biotechnology I*, New York: Springer-Verlag Berlin Heidelberg, 2000.

Galton, L.: *Hypothyroidism: The Unsuspected Illness,* New York, U.S.A: Cromwell, 1976.

Heslot, H. and Gaillardin, C.: *Molecular Biology and Genetic Engineering of Yeast*, CRC Press, Boca Raton, FL 1992.

Jacobson, M.: *Safe Food: Eating Wisely in a Risky World,* Washington, DC: Living Planet Press, 1991.

Jensen, Robert G.: *Handbook of Milk Composition*, San Diego: Academic Press, 1995.

John A, Marcell, *Handbook of Small Business Valuation Formulas and Rules of Thumb,* Valuation Press, 1993.

Koli. P A *: Dairy Development in India : Challenges Before Co-Operatives*, Shruti Pub, Delhi, 2007.

Krimsky, Sheldon: *Biotechnics and Society: The Rise of Industrial Genetics*, New York: Praeger Publishers, 1991.

Lata Bhattacharya: *Biochemistry of Nutrition*, Discovery, Delhi, 2010.

Martin Bauer: *Biotechnology 1996–2000: The Years of Controversy*, London: Science Museum Press, 2001.

Mowery, D.C.: *Technology and Wealth of Nations*, Stanford: Stanford University Press, 1992.

Nyholt, D.H.: *The Vitamin & Herb Guide,* Alberta, Canada: Global Health Ltd., 1992.

Paul B.: *Food Biotechnology in Ethical Perspective*, Aspen, CO: Aspen Publishers, 1997.

Purich, D.L.: *Enzyme Kinetics and Mechanism Part D, Developments in Enzyme Dynamics, Methods in Enzymology,* Academic Press, San Diego, CA, 1995.

R.B. Bushnell.: *Dry Cow Feeding and Management*, A Western Regional Extension Publication, 1979.

Ram Chand: *Decision-Making of Dairy Beneficiaries : Role of Aspiration, Motivation and Knowledge*, Om Pub, Delhi, 2010.

Ramakant Sharma : *Chemical and Microbiological Analysis of Milk and Milk Products,* International Book Distributing, 2006.

Rao, P. Venkateshwara: *Dairy Farm Business Management*, Biotech Books, Delhi, 2008.

Schulz, A. R.: *Enzyme Kinetics*, Cambridge University Press, Cambridge, 1994.

Selden, G.: *Body Electric: Electromagnetism and the Foundation of Life,* New York, U.S.A: William Morrow & Co., 1987.

Singhal, K K and D D Sharma: *Advances in Dairy Animal Production,* International Book Distribut, 2003.

Spreer, Edgar: *Milk and Dairy Product Technology*, Translated by Axel Mixa. New York: M. Dekker, 1998.

Susanna Hornig: *A Grain of Truth: The Media, the Public, and Biotechnology*, Lantham, MD: Rowman and Littlefield, 2001.

Thackray, Arnold: *Private Science: Biotechnology and the Rise of the Molecular Sciences,* Philadelphia: University of Pennsylvania Press, 1998.

Thompson, Paul B.: *Food Biotechnology in Ethical Perspective*, Aspen, CO: Aspen Publishers, 1997.

Tomar, S.K. and Gunjan Goel: *Applied Dairy and Food Microbiology,* Agrotech, 2005.

Venkateshwara Rao: *Dairy Farm Business Management*, Biotech Books, Delhi, 2008.

Wade, C.: *Helping Your Health with Enzymes,* New York, U.S.A: Universal-Award House, 1971.

Werbach M.R.: *Nutrition Influences on Mental Illness*, Third Line Press, 1999.

Yadav, D.S.: *Advances in Vegetable Production*, Researchco Book Centre, Delhi, 2005.

Index

G

H

I

L

M

P

Q

R

S

T

V

W

❑❑❑